Places of Possibility

Antipode Book Series

General Editor: Vinay Gidwani, University of Minnesota, USA
Like its parent journal, the Antipode Book Series reflects distinctive new developments in radical geography.
It publishes books in a variety of formats – from reference books to works of broad explication to titles that
develop and extend the scholarly research base – but the commitment is always the same: to contribute to
the praxis of a new and more just society.

Published

Places of Possibility: Property, Nature and Community Land Ownership
A. Fiona D. Mackenzie

The New Carbon Economy: Constitution, Governance and Contestation
Edited by Peter Newell, Max Boykoff and Emily Boyd

Capitalism and Conservation
Edited by Dan Brockington and Rosaleen Duffy

Spaces of Environmental Justice
Edited by Ryan Holifield, Michael Porter and Gordon Walker

The Point is to Change it: Geographies of Hope and Survival in an Age of Crisis
Edited by Noel Castree, Paul Chatterton, Nik Heynen, Wendy Larner and Melissa W. Wright

Privatization: Property and the Remaking of Nature-Society
Edited by Becky Mansfield

Practising Public Scholarship: Experiences and Possibilities Beyond the Academy
Edited by Katharyne Mitchell

Grounding Globalization: Labour in the Age of Insecurity
Edward Webster, Rob Lambert and Andries Bezuidenhout

Privatization: Property and the Remaking of Nature–Society Relations
Edited by Becky Mansfield

Decolonizing Development: Colonial Power and the Maya
Joel Wainwright

Cities of Whiteness
Wendy S. Shaw

Neoliberalization: States, Networks, Peoples
Edited by Kim England and Kevin Ward

The Dirty Work of Neoliberalism: Cleaners in the Global Economy
Edited by Luis L. M. Aguiar and Andrew Herod

David Harvey: A Critical Reader
Edited by Noel Castree and Derek Gregory

Working the Spaces of Neoliberalism: Activism, Professionalisation and Incorporation
Edited by Nina Laurie and Liz Bondi

Threads of Labour: Garment Industry Supply Chains from the Workers' Perspective
Edited by Angela Hale and Jane Wills

Life's Work: Geographies of Social Reproduction
Edited by Katharyne Mitchell, Sallie A. Marston and Cindi Katz

Redundant Masculinities? Employment Change and White Working Class Youth
Linda McDowell

Spaces of Neoliberalism
Edited by Neil Brenner and Nik Theodore

Space, Place and the New Labour Internationalism
Edited by Peter Waterman and Jane Wills

Forthcoming

Banking Across Boundaries: Placing Finance in Capitalism
Brett Christophers

Fat Bodies, Fat Spaces: Critical Geographies of Obesity
Rachel Colls and Bethan Evans

Gramscian Geographies: Space, Ecology, Politics
Edited by Michael Ekers, Gillian Hart, Stefan Kipfer and Alex Loftus

The Down-deep Delight of Democracy
Mark Purcell

Places of Possibility

Property, Nature and Community Land Ownership

A. Fiona D. Mackenzie

A John Wiley & Sons, Ltd., Publication

This edition first published 2013
© 2013 A. Fiona D. Mackenzie

Blackwell Publishing was acquired by John Wiley & Sons in February 2007. Blackwell's publishing program has been merged with Wiley's global Scientific, Technical, and Medical business to form Wiley-Blackwell.

Registered Office
John Wiley & Sons, Ltd, The Atrium, Southern Gate, Chichester, West Sussex, PO19 8SQ, UK

Editorial Offices
350 Main Street, Malden, MA 02148-5020, USA
9600 Garsington Road, Oxford, OX4 2DQ, UK
The Atrium, Southern Gate, Chichester, West Sussex, PO19 8SQ, UK

For details of our global editorial offices, for customer services, and for information about how to apply for permission to reuse the copyright material in this book please see our website at www.wiley.com/wiley-blackwell.

The right of A. Fiona D. Mackenzie to be identified as the author of this work has been asserted in accordance with the UK Copyright, Designs and Patents Act 1988.

Wiley also publishes its books in a variety of electronic formats. Some content that appears in print may not be available in electronic books.

Designations used by companies to distinguish their products are often claimed as trademarks. All brand names and product names used in this book are trade names, service marks, trademarks or registered trademarks of their respective owners. The publisher is not associated with any product or vendor mentioned in this book. This publication is designed to provide accurate and authoritative information in regard to the subject matter covered. It is sold on the understanding that the publisher is not engaged in rendering professional services. If professional advice or other expert assistance is required, the services of a competent professional should be sought.

Library of Congress Cataloging-in-Publication Data
Mackenzie, Fiona (A. Fiona D.)
    Places of possibility : property, nature and community land ownership /
    A. Fiona D. Mackenzie.
        p.   cm.
    Includes bibliographical references and index.
    ISBN 978-1-4051-9172-2 (cloth) – ISBN 978-1-4051-9171-5 (pbk.)
        1. Land reform–Scotland.   2. Land tenure–Scotland.   3. Commons–Scotland.
    4. Right of property–Scotland.
        HD1333.G72S2765 2012
        333.3'1411–dc23
                                                                            2012002563
A catalogue record for this book is available from the British Library.

Cover images: Traigh Mheilan beach, Glen Bhiogadal with the Cliseam, Urgha turbine, Ceann an Ora housing. Reproduced with kind permission © North Harris Trust
Cover design by www.cyandesign.co.uk

Set in 10.5/12.5pt Sabon by SPi Publisher Services, Pondicherry, India
Printed in Malaysia by Ho Printing (M) Sdn Bhd

1   2013

For Errol and Rowan

Contents

List of Maps

List of Photographs

List of Tables

Acknowledgements

This book is intended as a tribute to the many people on the Isle of Harris, Outer Hebrides, Scotland, with whom it has been my privilege to carry out research over the past 16 years. Initially, in the mid-1990s, this research focused on how those living on the island reworked what I would now refer to as their "subject positions" in the face of a multinational corporation's proposal to locate a massive superquarry in a National Scenic Area, within whose bounds the island lies. Subsequently, when the North Harris Estate, which comprises the north of the island, became the subject of community land purchase in 2003, the research focus changed to exploring how residents of North Harris, collectively, arc reworking ideas of property and nature in the search for more socially just and sustainable futures. It is this story – of how people in North Harris are in the process of reversing processes of dispossession that have taken different forms over hundreds of years – that is the subject of this book. It is a story that projects this place, so often written off historically as the periphery of the periphery, into the forefront of debates about how norms of global neoliberalization may be challenged in ways that create spaces for alternative, more socially just and sustainable possibilities.

The story could not have been told without the generous support over the years of the many people involved with the North Harris Trust, the community organization that owns the estate, whether in their capacity as directors of the trust, as employees or as members. My thanks go, first, to Calum MacKay, who, as Chair of the North Harris Trust, facilitated the research, supporting the extended discussions I have had with trust employees, my attendance at directors' meetings and making time in his busy schedule to talk. For the many, many, conversations I have had with them – topics ranging from the specifications of wind turbines, the particularities of the challenges of working with the land and the intricacies of housing policy to the profound differences which community land ownership makes – I owe an especial debt to David Cameron, Alistair Macleod and Duncan MacPherson. All three have been exceptionally generous with their time and without the detailed narration of their experience in setting up and working through a community land owning trust to effect more sustainable futures, the connections I have been able to explore through this book – between "property" and "nature" – would have been far less secure. To David Cameron, now Chair of the new organization, Community Land Scotland, as well as Chair of the North Harris Trading Company, the wholly owned subsidiary of the North Harris Trust, my thanks for sharing a vision of the future that makes visible an alternative to the historically embedded practices of social, cultural and economic erosion. To other directors of the trust, my thanks particularly to Cathie Bell Morrison, Barbara MacKay and Kenny Mackay. I would also like to record my thanks to Murdo Morrison, who has not lived to see the completion of the research but whose passion for the land invigorated every conversation I had with him. My thanks to David Wake and Robin Reid, both now working for the trust, whose insights, respectively, into renewable energy and the intricate interrelationship between people and the land/nature, have enriched this study. And my thanks to Mary Christine MacLennan and Diana MacLennan, also working for the trust, for sharing their understanding of crofting matters, and to Ann-Marie Hewitt for help with photographs. Others in North Harris whom I would like to thank include Ian Scarr Hall, Kenny Maclean, Innis Morrison and Fiona Morrison.

To Murdo MacKay, now chair of West Harris Crofting Trust, my thanks for explaining the intricacies of crofting law on numerous occasions. To Morag Munro, Councillor for Comhairle nan Eilean Siar/ Western Isles Council, a big thank you for sharing so many conversations about and beyond land reform. Thank you Morag Maclennan for help with Gaelic and Neil Campbell for discussion about the "business" side of things. Kenny Macleod provided valuable information about Harris Development Limited and Stephen Liddle detailed knowledge about the island's biodiversity.

On the Isle of Lewis, my thanks to Agnes Rennie, Carola Bell and Lisa MacLean of the Galson Trust. To John Randall of Pàirc Trust, particular thanks for discussions about the community's ongoing struggles of gaining ownership of the land. Thank you, too, to Huw Francis of Stòras Uibhist, for the time taken to talk about that community's vision for the future. From further afield, thank you to Angela Williams of the Knoydart Foundation. Rona Womersley and Kathleen Maclennan of Community Energy Scotland provided detailed information about community owned wind farms and Mairi Mackinnon, planning officer at the Comhairle, facilitated my access to documents pertaining to wind farm applications, large and small. My thanks for conversations with Patrick Krause and Donald Murdie of the Scottish Crofting Federation, with John Watt, Sandra Holmes and Donnie MacKay of Highlands and Islands Enterprise, with Nigel Hawkins and Mick Blunt of the John Muir Trust and with Will Boyd-Wallis of the Cairngorms National Park.

I have benefited immeasurably from discussions with colleagues in the academic community. Most recently, my thanks to Martin Price, Centre for Mountain Studies, Perth College, University of the Highlands and Islands, for providing a base during my last sabbatical and discussions with Jim Hunter, Calum Macleod, Andy Wightman and Maria Scholten during that period. For the previous sabbatical spent at the Institute of Governance, University of Edinburgh, my thanks to David McCrone, Margaret MacPherson and Lindsay Adams for a warm welcome. At the University of Edinburgh, my thanks also to Ewen Cameron and Lynn Jamieson. And, during my years of involvement with the University of Aberdeen, my particular thanks to John Bryden, Mark Shucksmith and Fraser MacDonald of the Arkleton Institute. At Carleton University, I would like to express my deep appreciation for their support to John Osborne, Dean, Faculty of Arts and Social Sciences, and Mike Brklacich, Chair, Department of Geography and Environmental Studies, and colleagues Simon Dalby and Trish Ballamingie. My thanks also for stimulating discussions with postgraduate students with whom it has been my privilege to work, particularly Brian Egan, Andrew Baldwin, Sherrill Johnson and Abra Adamo. And, to Justin Stefanik and France-Lise Colin, my thanks for both good discussions and work as research assistants. For drafting maps, a thank you to Adèle Michon at Carleton University and Mick Ashworth, Glasgow, and for things technical, a thank you to Dan Patterson. I would like to record my appreciation to the Social Sciences and Humanities Research Council of Canada for funding the research through a series of grants.

I would also like to thank Rachel Pain, Editor of the Antipode Series in which this book is published, and Jacqueline Scott, Isobel Bainton and Eunice Tan of Wiley-Blackwell, for their support and encouragement during the book's long gestation. This book is the culmination of personal research carried out over many years. The ideas on which it is based have developed as research has progressed. Earlier analysis of some of the material has appeared in sections of the following publications: 'S Leinn Fhèin Fearann (The Land is Ours): re-claiming land, re-creating community, North Harris, Outer Hebrides, Scotland, *Environment and Planning D: Society and Space*, 2006, 24:577–598; A working land: crofting communities, place and the politics of the possible in post-Land Reform Scotland, *Transactions of the Institute of British Geographers*, 2006, NS 31: 383–398; Working the Wind: Land-owning Community Trusts and the Decolonization of Nature, *Scottish Affairs*, 2009, 66: 44–64; and A common claim: community land ownership in the Outer Hebrides, Scotland, *International Journal of the Commons*, 2010, 4 (1): 319–344.

A thank you too to my father, whose warm welcome to his home near Perth as I was en route to or from the Isle of Harris I could always count on. And to those who have sustained me over the years through long walks in the hills – 'bagging' Munros – my son Errol and daughter Rowan, Nick Spedding, Sandy Mather and George Yeomans, a special thanks for keeping my feet firmly on the ground whatever possibilities about the land my mind might have been imagining.

1

Placing Possibility

The Community Land Conference, 29–30 September 2009, Isle of Harris, Scotland

On 29 and 30 September 2009, representatives of the 20 largest community land owning trusts, responsible for the management of over 400 000 acres of land in Scotland's Highlands and Islands, met in the Harris Hotel, Tairbeart, Isle of Harris, in the Outer Hebrides, to discuss how to take forward community-centred land reform.[1] Delegates at the Community Land Conference shared a concern that despite the well-publicized successes of the community land ownership movement – underway since the historically unprecedented purchase by the Assynt Crofters' Trust of the North Lochinver Estate in 1993 and supported, a decade later, by the Land Reform (Scotland) Act 2003 – momentum had been lost (Table 1.1).[2] A beginning had been made in reversing the extraordinarily skewed distribution of land in Scotland,[3] but the process of democratizing land ownership was far from complete. Funding streams that were essential to communities seeking to purchase land that had been held as private estates for hundreds of years, frequently by absentee landlords, were no longer guaranteed, and land reform itself seemed to have slipped from the Scottish Government's agenda. The task for delegates was to consider how to take the land reform process forward such that other communities could search out its political possibilities.

Places of Possibility: Property, Nature and Community Land Ownership,
First Edition. A. Fiona D. Mackenzie.
© 2013 A. Fiona D. Mackenzie. Published 2013 by Blackwell Publishing Ltd.

Table 1.1 Community land ownership in Scotland, 2010 (Property over 2000 acres).

Property	Owner	Acres	Hectares	Date of acquisition
West Harris Estate	West Harris Crofting Trust	16255	6578	2010
Galson Estate	Urras Oigreachd Ghabsainn	56000	22662	2007
South Uist Estate	Stòras Uibhist	93000	37636	2006
Glencanisp and Drumrunie Estates	Assynt Foundation	44578	18047	2005
North Harris Estate	North Harris Trust	55000	22267	2003[i]
Gigha	The Isle of Gigha Heritage Trust	3694	1495	2002
Little Assynt Estate	Culag Community Woodland Trust	2940	1190	2000
Forest of Birse	Birse Community Trust	9000	3642	1999
Knoydart Estate	Knoydart Foundation	16771	6787	1997
Eigg	Isle of Eigg Heritage Trust	7263	2939	1997
Melness	Melness Crofters' Estate	12522	5067	1995
Borve and Annishader Estate	Borve and Annishader Township	4502	1822	1993
North Assynt Estate	Assynt Crofters' Trust	21132	8552	1993
Stornoway Estate	Stornoway Trust	69400	28085	1923

[i] In 2006, the Loch Seaforth Estate was added to the North Harris Estate, bringing the total acreage to 62500 (25304 ha).
Source: Adapted from Wightman, 2010: 150.

In his introductory keynote address to the conference, renowned historian James Hunter reminded the audience of John McGrath's immensely popular play *The Cheviot, the Stag and the Black, Black Oil*, performed throughout the Highlands and Islands in 1973. "For all its entertaining format" – the play was staged in the vernacular idiom of a ceilidh – Hunter emphasized its serious message, spoken in the final scene by the full cast: "The people do not own the land. The people do not control the land" (Hunter, 2009: 1). He expanded,

> Whether by clearing lairds, by the absentee owners of sporting estates or by the multinational corporations then beginning to be involved with North Sea oil, the resident population of the Highlands and Islands, or so McGrath contended, had been denied any jurisdiction over their area's natural assets – just as they'd been deprived, McGrath argued, of any

substantial share of the profits and revenues deriving from the commercial exploitation of those assets (Hunter, 2009: 1).[4]

That was now changing. Since that "revolutionary moment" in the summer of 1992 when the Assynt crofters decided to bid for the North Lochinver Estate (now the North Assynt Estate), placed on the market by the liquidators of Scandinavian Property Services Ltd, which had owned the land since 1989 (see MacAskill, 1999), people had begun to question what had previously been taken for granted – namely, "that private estates would forever be bought and sold without reference to, or interference from, the people living on them" (Hunter, 2009: 3).[5]

Hunter summarized the remarkable successes of community land ownership. In material terms, achievements included the provision of new housing and the upgrading of existing housing, increases in population, the establishment of new businesses and job opportunities, "more environmentally-sensitive management of a whole range of natural habitats" and renewable energy schemes (Hunter, 2009: 4). Less easy to measure, he noted, but "hugely important", was evidence of growing "self-esteem" and "self-confidence" (Hunter, 2009: 4). He challenged those present to raise the political profile of community land ownership in Scotland and to lobby the government for a renewed commitment to land reform. This government, minority though it might be, was after all formed by the Scottish National Party, a party that, while "far from power" in 1973, had invited McGrath to stage a performance of *The Cheviot, the Stag and the Black, Black Oil* at their annual conference (Hunter, 2009: 10).

Unanimously, representatives of the community trusts present adopted a series of proposals designed to further community land ownership in the Highlands and Islands.[6] They mandated the working group of three who had organized the conference to take these proposals forward to the Scottish Government and to lead an investigation into the possibility of more formal political representation of community land owning groups. The name C-20, lightly quipped at the first meeting in Inverness of a steering group where the idea to hold a conference was first mooted, provided an interim collective identifier for this purpose. With a measure of irony, the name's verbal resonance with the G-20 conjures an identification with the global, but simultaneously undercuts any claim to a global defined through the reductionist deliberations of finance ministers and the governors of central banks from 20 economies. The name C-20 suggested a claim to an alternative, place-based and more generous politics to that of a neoliberal imaginary.

4   PLACES OF POSSIBILITY

Places of possibility

This is a book about the land reform process in Scotland's Highlands
and Islands, which so inspired those present at the conference to seek
ways of ensuring its continuation. It is also an engagement with the
broader questions – of property, nature and neoliberalization – through
which this struggle is constituted. The book considers how community
land ownership opens up the political terrain of particular places through
the reconfiguration of practices of property and of nature to more
socially just and sustainable possibilities than those prefigured through
prevailing norms of neoliberal practice, specifically enclosure and
privatization. More precisely, it explores how the complicated and
contingent process of "commoning" the land through community
ownership troubles binaries – of public/private and nature/culture – and
through these disruptions creates a space/place where neoliberalism's
normalizing practices are countered. The islands of the Outer Hebrides
(or the Western Isles, as these islands are also known) are shown to be
places of possibility where norms that had previously confined political
possibility are now unsettled and new imaginaries configured.

Within Scotland, at "the cusp", globally, of "community-centric" land
reform (Bryden and Geisler, 2007), the Outer Hebrides provide an exceptionally rich area in which to explore the political possibilities that are
created as ideas of property and nature are reworked through community
land ownership.[7] First, after a lengthy history of dispossession – of the
enclosure and privatization of rights to the land associated both with the
Clearances of the eighteenth and nineteenth centuries and, more recently,
of the collapse of fish stocks, the decline in the price of sheep and the
vicissitudes of the (primarily Norwegian) corporately controlled fish
farms – it is these islands that are at the forefront of the land reform
movement. Well over one-third of the land in the Outer Hebrides is now
in community ownership and over two-thirds of the population resides
on community owned estates (West Highland Free Press (*WHFP*), 19
January 2007: 1) (Map 1.1; Table 1.2). Further community purchase of
land is the subject of ongoing discussion, Pàirc Estate on the Isle of Lewis
being one example (Chapter 4) and the island of Scalpay, Harris, another.
Places that had long been considered "peripheral" to economic life in
Scotland are now, ironically, at the forefront of initiatives to achieve more
sustainable futures – economically, ecologically, socially and culturally.

Second, it is in these islands – arguably the windiest and certainly
among the "wildest" areas in Scotland – that struggles over nature are
acute. On the one hand, the Outer Hebrides are caught up in globalizing

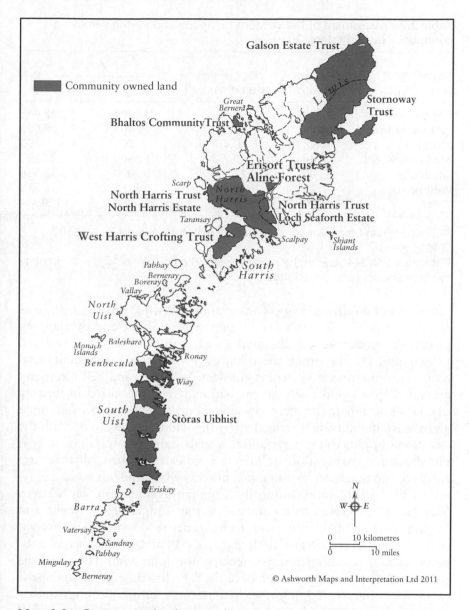

Map 1.1 Community land ownership in the Outer Hebrides. © Ashworth Maps and Interpretation Ltd 2011. Contains Ordnance Survey data © Crown copyright and database right 2011.

Table 1.2 Assessment of the Western Isles population living within a community owned estate, 2010.

Community owned estate	Approx. population	% of Western Isles population[i]	Area[ii]	
			Acres	Hectares
Galson	2139	8.1	56000	22662
Stornoway Trust	12015	45.3	69400	28085
Bhaltos	98	0.4	1705	690
North Harris	704	2.7	62500	25293
South Uist	3200	12.1	93000	37636
West Harris	123	0.5	16255	6578
Total	**18279**	**69**	**298860**	**120944**

[i] Comhairle nan Eilean Siar Local Authority Area's total population – 26502 (2001 Census).
[ii] There are 758844 acres in the Western Isles (38.6 per cent of land is community owned, excluding Ath Linne forests).

discourses of biodiversity and conservation of "wild" areas. Large areas of the land (and adjacent seas) are subject to multiple and overlapping protective environmental designations of national, UK and European provenance. On the other, the islands are the site of community and corporate initiatives to harness the wind's energy and sell electricity generated from wind farms to the national grid, mobilized in turn by discourses of climate change, renewable energy targets and local/national/ global sustainability. In a struggle over the wind that has to do with the assertion of rights to property and to a sustainable future, community is pitted against corporation or private syndicate. In turn, all three are party to a complex decision-making process where the other main players are the local planning authority (Comhairle nan Eilean Siar/Western Isles Council, referred to commonly as the Comhairle by Gaelic and English speakers alike), the government agency responsible for conservation matters (Scottish Natural Heritage, SNH), and environmental nongovernmental organizations, particularly the John Muir Trust and the Royal Society for the Protection of Birds. Who has the right to define the meanings of nature may, in these circumstances, be hotly contested.

Third, these islands share common challenges of social, cultural and economic fragility and thus hopes that community ownership of the land can reverse this situation are high. A recent study commissioned by Comhairle nan Eilean Siar/The Western Isles Council drew attention to alarming population trends (HallAitken, 2007). Overall, in these islands, data indicate that the population fell by 40 per cent between 1901 and

2001, with the steepest declines in Harris, the Uists and Barra (HallAitken, 2007: 1). The population of Harris, for instance, fell by 60 per cent over this time period, from a total of 5271 in 1901 to 2120 in 2001 (HallAitken, 2007: 12). Between 1991 and 2001, a 12 per cent decline was recorded (Bryden et al., 2008: 25). More recent data for the islands as a whole suggest a small increase in population of 1 per cent between 2003 and 2005, with an uneven spread among age groups and geographically (HallAitken, 2007: 13). The largest increase (10 per cent) was in the age group 55–59, characterized by the report as "lifestyle in-migration"; disturbingly, over the same time period, the number of children under the age of 15 continued to decline (HallAitken, 2007: 13, 19). The ageing of the population is evident – in 2001, 35 per cent of the population of Harris and Scalpay were aged 60 or over, a figure that compares with 26 per cent for the Outer Hebrides as a whole (Bryden et al., 2008: 25). On Harris itself, there has been a reversal of the downward population spiral since 2003 and, between 2003 and 2005, an increase of 10 per cent in the primary school roll (HallAitken, 2007: 17, 16).

Outmigration, particularly by women, who leave in greater numbers than men, is singled out as significant in explaining population trends (HallAitken, 2007: 2). The report cites as the main reasons for outmigration the limited number of job opportunities, particularly those that fall into the category of "skilled", the search for further education and then employment on the Scottish mainland, and the lack of affordable housing (HallAitken, 2007: 2). The acquisition of houses as second or holiday homes, as well as demand from older people (often retirees) moving to the islands, notes the report, pushes up prices such that the young, in particular, can no longer afford to buy (HallAitken, 2007: 30). Against a Scottish average of 1.3 per cent of the housing stock in second or holiday homes, the figure for the Outer Hebrides is 7.2 per cent; in West Harris, this figure rises to 17 per cent (HallAitken, 2007: 30–31). The report lists five "factors" as critical in the reversal of these trends and in the creation of "sustainable communities": sustainable employment, private-sector led economic diversity, the provision of affordable housing, "self-determination" and "clean energy" (HallAitken, 2007: 4–5), issues to which I return later. It is also the case that the islands face the threat of cultural loss. As the "heartland" of the Gàidhealtachd, it is here that is found the highest proportion of Gaelic speakers in Scotland, but this number is declining. On Harris, 81.7 per cent of the population (1861 people) spoke Gaelic in 1991; this was reduced to 69.9 per cent (1447 people) in 2001 (Bryden et al., 2008: Appendix 5: 21).[8]

Among the several communities that now own land in the Outer Hebrides, I place particular attention on Urras Ceann a Tuath na

Photograph 1.1   Cairn marking the North Harris Trust.
Source: author.

Hearadh/the North Harris Trust (NHT), which bought the 55 000 acre (22 267 ha) North Harris Estate from Jonathan Bulmer in 2003, to which was added the 7500 acre (3036 ha) Loch Seaforth Estate in 2006 (NHT, 2007: 6). In total, the NHT owns 62 500 acres (25 304 ha) of land, 29 300 acres (11 862 ha) (46.9 per cent) of which are under crofting tenure and 33 200 acres (13 441 ha) (53.1 per cent) of which lie outwith crofting tenure (NHT, 2007: 18). The population is about 700, the majority living in Tarbairt; about 250 live in the various crofting

Photograph 1.2   The hills of North Harris from Ben Luskentyre in South Harris. *Source*: author.

townships located along the coast (NHT, 2007: 6). In addition to some income from crofting, the main sources of employment are public service, fishing, fish farming and construction (NHT, 2007: 28). In addition, tourism is a major income earner (NHT, 2007: 28).

   In one sense, the strongest methodological reason for this choice is that, of the three largest land owning community trusts in the islands (the other two being Urras Oighreachd Ghabhsainn/the Galson Estate Trust, Isle of Lewis, and Stòras Uibhist/South Uist Estate, South Uist, Eriskay and Benbecula), the North Harris Trust is the oldest (Photograph 1.1). It is thus here that the negotiation of the political possibilities of community land ownership has occurred over the longest period of time. It is also the case that it is on the North Harris Estate that some of the most visible struggles over nature – as "the wild" and with respect to the wind – have occurred (Photograph 1.2). However, the more immediate rationale for this choice is that I have carried out research in the Isle of Harris since 1995 – a time when a multinational corporation's proposal for planning permission for a superquarry at Lingerbay, in the south of the island, was the subject of a public inquiry (Barton, 1996; Owens and Cowell, 1996; Mackenzie, 1998b; McIntosh, 2004). I thus have extensive research experience in this area. An agreement to conduct research with the North Harris Trust grew from this earlier and ongoing research and was negotiated in 2002, at which time the property was first placed on the market (Mackenzie, 2006a).

This is, then, a case study. Through the particularities of in-depth qualitative research, I document and then analyse the ways through which community land owning trusts rework ideas of property and nature such that their political possibilities are made visible. As a case study, and in line with the direction of thinking outlined in the next section of this chapter, the research seeks not to provide the grounds for generalization – for the impact of or resistance to processes of neoliberalization. Instead, it uses the particulars of a place as "windows into constitutive processes, and a means for reconfiguring understandings and practices" (Hart, 2004: 97). Specifically, it searches for the ways through which normalization – produced by totalizing narratives of property and nature – is disrupted. It does so, primarily, by questioning the "self-evidence" or "givenness" of categories – of property and nature, exploring how these categories are produced materially and discursively in the interest of a particular politics. As an "evidence-based" or "contextual" study – Castree (2008a) uses the two terms interchangeably – of the political possibilities of community land ownership, its broader import lies in its identification of political openings where the "inevitability" of neoliberalization and, specifically, practices of enclosure and privatization are troubled, and new, more hopeful, futures may be imagined.

With the intent of making visible the intricacies – the complexities, the contradictions and the contingencies – that are part and parcel of the process of negotiating the political possibilities of particular places, I have employed a range of research methods. These have involved, first, participation in community-initiated events, North Harris Trust directors' meetings, community-led or community-focused workshops (for example, on housing and renewable energy), relevant meetings of the Comhairle nan Eilean Siar (for example, when proposals for wind farms were being discussed) and annual gatherings of the Scottish Crofting Federation. In addition, I have participated in meetings of the Cross Party Group on Rural Policy and the Cross Party Group on Crofting at the Scottish Parliament. Second, I have carried out semi-structured or in-depth interviewing, on a regular basis, with North Harris Trust directors, employees and members who have responsibility for specific initiatives since 2003. I have also carried out interviews with key informants from the Galson Trust, Stòras Uibhist, the Knoydart Foundation, Comhairle nan Eilean Siar, Scottish Natural Heritage, the Scottish Crofting Federation, the Cairngorms National Park Authority and the John Muir Trust. Third, I rely on primary documentary material from the organizations I have just identified and from the Scottish Government. With respect to the Comhairle, documents pertaining to applications for planning permission for wind farms from communities, corporations and individuals have included letters of support or objection from the public as well as from statutory authorities such as Scottish Natural Heritage and the Scottish Environmental Protection Agency. Fourth, I also draw on newspaper reports and letters to the editor, particularly those from the two main local papers, *The West Highland Free Press* and *The Stornoway Gazette*.

Fifth, in places, I call on literature, particularly poetry, from the Gàidhealtachd, as the evidential base for exploring both the historical beginnings of contemporary struggles over, and current thinking about, land and nature. Poets, writes John MacInnes (2006: 3), "are the spokes[people] of Gaelic society. ... It is not a different awareness so much as a difference in artistic convention that makes the Gaelic poet concern him [or her]self with the national dimensions of a given issue". It is Gaelic poetry, emphasizes Donald Meek (1976: 309–310), that provides the historical evidence for people's experience of the Clearances and the "Land Agitation" that followed. It was song in particular that was "until recently", he recalls (Meek, 1976: 310), "the principal medium of popular journalism", and "Gaelic society had for long afforded considerable prestige to the poet as a commentator on current ideas and events". For twentieth-century Lewis-born poet Iain Crichton Smith, songs and poems provide "a kind of history lived on the bone rather than an intellectual creation" (cited by Hunter, 1995a: 26). Congruent with Edward Said's (1994) broader theorization of "a culture of resistance" against imperial rule, Hunter (1995a: 26–28) argues that the collective poetical archive of a people – their songs, poems and stories – provides a more reliable barometer of local experience than the dispassionate reconstructions of the past by historians where this is unmediated by postcolonial theorization.[9] It is also the case that, limited though analysis remains, women poets provide a means of balancing what are undoubtedly male-centric productions of history. Despite social conventions of the times, "Highland women", writes Michael Newton (2009: 158), "have enjoyed surprisingly prominent roles in the creation and transmission of Gaelic literature in nearly every century".

As a final methodological note, in analysing the material cited, I have been concerned to focus both on words – written and oral – and on the visual. The search for a new politics of the possible with community land ownership is bound up with troubling the visual as well as the verbal order through which property and nature are *"given* to be seen, how [they are] *'shown'* to knowledge or to power" (Rajchman, 1991: 69–70, emphasis his). A counternarrative questions the "exclusionary geographies" (Gregory, 2000: 314) through which the land was normalized, re-mapping it in ways that are visually at odds with narratives of a sporting estate empty of people or a place of "wildness" that must be protected from people. It replaces this colonizing optic – "the sole scopic regime" (Jay, 1994: 589) of neoliberalization – with one that is more complex and contingent or, in Martin Jay's (1994: 592) language, "polyscopic".

Opening theoretical places

The theoretical – and methodological – initial reference point for probing the "givenness" of property and nature is Michel Foucault's (1979, 1985, 2007) writing on the "workings" or the "how" of power in so far as it concerns the production of norms, processes of normalization and their reversibility. This "analytics" (Foucault, 1985: 82) traces how power is exercised and resisted. "Power produces", writes Foucault (1979: 194), "it produces reality; it produces domains of objects and rituals of truth". It proceeds, as he shows in *Discipline and Punish* (1979) and *The History of Sexuality* (1985), through the creation of norms which, since the eighteenth century, have served to regulate society. "Like surveillance and with it", he asserts, "normalization becomes one of the great instruments of power at the end of the classical age" (Foucault, 1979: 184). It homogenizes but at the same time "individualizes by making it possible to measure gaps, to determine levels, to fix specialities and to render the differences useful by fitting them one to another" (Foucault, 1979: 184). It creates abnormalities – in Foucault's work, for example, of criminality or sexuality – which must then be "treated". It operates as a "political technology", removing from the political domain something that is basically "a political problem", reinventing it as a politically inert "technical problem" (Dreyfus and Rabinow, 1983: 196).

As John Rajchman (1991) explains in his analysis of "Foucault's Art of Seeing", visuality is internal to the process of normalization. There is, he writes, with reference to *Discipline and Punish* (Foucault, 1979), an "interconnection between seeing, doing, and practical self-evidence" (Rajchman, 1991: 79). Normality "becomes 'visible' through an expanding network of practices" (Rajchman, 1991: 79). Visibility is thus "one of the great 'self-evidences' of the workings of power" (Rajchman,

1991: 84). "Power conceals itself", he continues, "by visualizing itself. Its workings become acceptable because one sees of it only what it lets one see, only what it makes visible" (Rajchman, 1991: 84).

The process of normalization does not go uncontested. In "What is critique?", Foucault (2007: 66) emphasizes that power may not be "understood as domination, as mastery, as a fundamental given, a unique principle, explanation or irreducible law". Rather, it has to be "considered in relation to a field of interactions, contemplated in a relationship which cannot be dissociated from forms of knowledge" (Foucault, 2007: 66). He continues, "One has to think about it in such a way as to see how it is associated with a domain of possibility and consequently, of reversibility, of possible reversal" (Foucault, 2007: 66). Reversibility and the creation of what Julie Katherine Gibson-Graham (2006: xiv) calls a "politics of possibility" have to do with disturbing "the acceptability of the system", of disrupting norms (Foucault, 2007: 61).

Taking Foucault's theorization of the norm into the realm of feminist politics, Judith Butler (2004) traces how new political possibilities are created when prevailing norms of gender and what counts as "human" are "undone". Particular ways of "doing" gender, she shows – as drag, butch, femme, transgender and transsexual persons – trouble prevailing norms of what counts as human, which depend on the binary male/ female. What are the consequences, she asks, "if the very categories of the human [exclude] those who should be described and sheltered within its terms?" (Butler, 2004: 36). "If we take the field of the human for granted", she continues, "then we fail to think critically and ethically about the consequential ways that the human being is produced, reproduced and deproduced" (Butler, 2004: 36). While disturbing the category gender is not the focus of this book's attention, Butler's ideas provide an analytic template for undoing the categories property and nature and tracing the ways through which they are both constituted through the binaries public/private and nature/culture and unsettled when these binaries are questioned – by a commoning of the land.

In line with Foucault's reasoning, Butler (2004) argues that a politics of new possibilities proceeds by opening up or "undoing" such categories as human and gender, by interrupting the norms – "the settled knowledge and knowable reality" (Butler, 2004: 27) – through which the notion of human and the binary of man and woman have been constituted and, through this opening, to imagine a new reality. Drag, butch, femme, transgender and transsexual persons, she proposes, do precisely that. "They make us not only question what is real, and what 'must' be, but they also show us how the norms that govern contemporary notions of reality can be questioned and how new modes of reality can become instituted" (Butler, 2004: 29). They "rattle" norms, "display their instability, and ... open them to resignification" (Butler, 2004: 28). They show how what it means to be human or of the gender man or woman is always "in process, underway, unfulfilled" (Butler, 2004: 37).

For the analyst, this means that it is necessary both to trace how a particular "nexus of power and knowledge" works "to constitute a more or less systematic way of ordering the world with its own 'conditions of the acceptability of a system'" (Butler, 2004: 215–216, with citations from Foucault) and "to track the way in which that field [of intelligible things] meets its breaking point, the moments of its discontinuities, and the sites where it fails to constitute the intelligibility it promises" (Butler, 2004: 215–216, drawing on Foucault). "[O]ne looks *both* for the conditions by which the object field is constituted", Butler (2004: 27, emphasis in original) argues, "and for *the limits* of those conditions". The limits, she continues, are found "where the reproducibility of the conditions is not secure, the site where conditions are contingent, transformable" (Butler, 2004: 27). It is this contingency and transformability, I suggest in this book, that are found where categories of property and of nature are opened up through community ownership to new political possibilities.

Gibson-Graham (2006) develops the idea of a politics of possibility or what William Connolly (1999: 57, cited by Gibson-Graham, 2006: xxiii) terms "a politics of becoming" by "queering" capitalism. She traces the interruptions – the openings – in the economy to trouble the norms through which capitalism is produced and how, through a process of economic normalization – where what counts as "economic" is defined through a "capitalocentric" imaginary – subjects are themselves "normalized" (Gibson-Graham, 2006: 25). The dislocation of "the unity and hegemony of neoliberal global capitalist economic discourse", she writes (2006: 56), is carried out "through a proliferative queering of the economic landscape and the construction of a new language of economic diversity". While my focus in this book is not on the economy *per se*, Gibson-Graham's ideas about a politics of possibility which she develops through the idea of a community economy and her use of Foucault's (1990) writing on "modes of subjectivation" and "practices of the self" are useful to take forward in the pursuit of the new politics of property and nature that, I argue, is instigated through community land ownership.

To turn first to the idea of a "diverse economy", Gibson-Graham (2006: 55) draws from such experiences as those of Mondragón in the

Basque region of Spain and her own research in the Latrobe Valley in Victoria, Australia, the Connecticut Valley in the northeastern United States of America and the Jagna Municipality in Bohol, southern Philippines, to unsettle the idea of capitalism as the sole economic signifier, as "extradiscursive, as the ultimate real and natural form of economy", effectively evacuating all other economic possibilities. Her counterhegemonic politics depends on "liberat[ing]" alternative economic languages from their "discursive subordination" (Gibson-Graham, 2006: 57) and creating a language of economic diversity – "a less capitalocentric, more inclusive, more differentiated language of economy" (Gibson-Graham, 2006: 2) that expands the idea of market exchange and includes alternative and nonmarket transactions (Gibson-Graham, 2006: Chapter 3). This language, she writes in the conclusion to A *Postcapitalist Politics*, is "our principal technology for 'repoliticizing the economy'" (2006: 195). It provides "a fragmentary and incoherent starting place" for the production of a new narrative of "a diverse (becoming community) economy" (2006: 195) or what, in a later work, is referred to as "a new econo-sociality" (Gibson-Graham and Roelvink, 2010: 330).

Gibson-Graham's research demonstrates that displacing subscription to a capitalist economic discourse may not be an easy task, even where the consequences of economic failure are as visible and devastating as in Argentina in the crisis of 2001. Members of the unemployed workers' movement who took over empty factories, she writes, had to do battle not with the state or capital but with "their own subjectivities" (Gibson-Graham, 2006: xxxv). They had to reject "a long-standing sense of self and mode of being in the world [as workers, rather than managers or entrepreneurs], while simultaneously cultivating new forms of sociability, visions of happiness, and economic capacities" (Gibson-Graham, 2006: xxxv). Gibson-Graham extends this discussion through the details of her work with the Community Partnering Project in the Latrobe Valley, Australia. She shows how their search for "openings and possibilities" at a time of massive unemployment caused by privatization and retrenchment in the electricity and mining industries began by unsettling "the naturalness" of the "Economy". By deconstructing and making visible the discourses through which the Economy was made "thinkable and manipulable" in the valley, their collaborative research project created the discursive grounds for an alternative. Privatization and retrenchment in the case of the Latrobe Valley or factory closures in the case of Argentina, she suggests, provided "a moment of interruption in ritualized practices of regional economic subjection" (Gibson-Graham,

2006: 25) and thus a moment where the process of rethinking what it means to be an economic subject might begin.

Resubjectivation – the process through which people reposition themselves *vis à vis* the economy – depends on new "practices of the self" (the phrase is Foucault's, 1990: 28).[10] In Gibson-Graham's (2006) work, this has to do with the performance of subjectivities other than those singularly and exclusively defined through a capitalist economy. For participants in a discussion group carried out in the context of research in the Latrobe Valley, it meant a refusal to position themselves as "subjects of insufficiency who need[ed] proper economic identities in the formal economy" (Gibson-Graham's, 2006: 143). The research tried to create spaces where people could position themselves as part of a more plural, diverse and "generous" economy than they had previously envisioned. One outcome was what Gibson-Graham (2006: 15) refers to as "the awakening of a communal subjectivity", one where, through recognition of their interdependence and the (already) rich diversity of their "economic" lives, people began to build a community economy.

In contrast to "the structurally configured Economy with its regularities and lawful relationships", the community economy is not a "blueprint" but "an unmapped and uncertain terrain that calls forth exploratory conversation and political/ethical acts of decision" (Gibson-Graham, 2006: 166). It is a place where the meanings of the economic and of community are never foreclosed but are kept open to "continual resignification" (Gibson-Graham, 2006: 98). It relies on a notion of "community" as constantly "a-doing", as always in the process of "becoming", or, in Linda Singer's (1991: 125, also cited by Gibson-Graham, 2006: 99) words, as "a call of something other than presence" rather than as "a referential sign". It implies a collective identity that is created in and through specific circumstances and in particular places, "a becoming in common" rather than "a common being" (the last phrase is Jean-Luc Nancy's, 1991: 4).[11]

Place in Gibson-Graham's conceptualization of a community economy is open, "not fully yoked into a system of meaning ...; it [is] the aspect of every site that exists as potentiality. ... It is the unmapped and unmoored that allows for new moorings and mappings" (Gibson-Graham, 2006: xxxiii). "Place, like the subject", she continues (Gibson-Graham, 2006: xxxiii), "is the site of becoming, the opening for politics". As such, places are not seen as simply "victims" of neoliberal globalization; nor are they necessarily "always politically defensible redoubts *against* the global", to draw from Doreen Massey (2005: 101, emphasis hers). Understood relationally as "criss-crossings in the wider power-geometries that constitute both themselves and the 'global'" (Massey, 2005: 101), places may provide sites where "the inevitability" (Massey, 2005: 103) or "normative insistence" (Massey, 2000: 283) of neoliberal globalization may be questioned and a narrative of political possibility kindled. Its "progressive" possibilities, Massey (2009: 415) states in a recent conversation, have to do with critically examining the meanings of "the local" and how the local relates with other places – in her words, "a politics of place beyond place" (Massey, 2009: 415).[12]

"Undoing" property/nature

I draw on these ideas of norms, processes of normalization and the political possibilities that are created through their disruption in order to position a discussion of community land ownership in relation to contemporary debates about property, nature and neoliberalism. My specific concern is to contribute to a literature defined broadly in terms of poststructural political ecology that has documented, with increasingly analytical rigour, the social and environmental injustices produced (if not ubiquitously, then certainly overwhelmingly) through processes of neoliberalization, but in which counternarratives are, with important exceptions, far less visible.[13] There has been, as Noel Castree (2008b: 168) remarks, at least with respect to research on "neoliberalizing nature", more focus on "understanding actualities" than on tracing "potentialities". Through an analysis of the principles and practices of community land owning trusts, primarily in the Outer Hebrides, my intention, through the detailed documentation of the potentialities of a particular place, is to further a geography of hope that is suggested by Castree's comments.

While its intellectual underpinnings go back a great deal further (Harvey, 2005), neoliberalism as a political economic project has gained ascendancy since the early 1970s as both institutional practice and hegemonic discourse. Its critics point out that, as Keynesian modes of social regulation were increasingly called into question at a time of lowering rates of profitability in the largest capitalist economies (Heynen *et al.*, 2007b: 6), neoliberal policies were adopted "to expand opportunities for capital investment and accumulation by re-working state-market-civil society relations to allow for the stretching and deepening of commodity production, circulation and exchange" (Heynen *et al.*, 2007b: 10). Governments employed a series of measures that supported the further penetration of the market, economically, socially, culturally, politically and environmentally – a process that has become known as neoliberalization. The well-rehearsed litany of measures includes more regressive forms of state taxation, cutbacks in social services, the privatization of other services for which the state previously had responsibility, the "reinforcement and extension of private, exclusive, and individuated property rights", trade liberalization (albeit, as Heynen *et al.*, 2007b: 6, remark, in contradictory ways, as subscription to an ideology of free trade was pitted against particular political interests), "workfare" and other efforts to "discipline" workers and civil servants together with the "deregulation and re-regulation of labor markets" (Heynen *et al.*, 2007b: 6), decentralization of government and the conscription of civil society organizations to plug the vacuum in social support created through state cutbacks, and "the restructuring of state regulatory apparatuses" such that significant authority for environmental matters, for example, passes to private and corporate institutions (Heynen *et al.*, 2007b: 6).

While this process of neoliberalization has, as Jamie Peck and Adam Tickell (2002: 384), point out, been "neither monolithic in form nor universal in effect", they argue that a broad trend may be discerned between those measures that they characterize as "roll-back" neoliberalism in the 1980s and the more recent "roll-out" neoliberalism. Whereas "roll-back" neoliberalism might be seen as focusing on "the active *destruction and discreditation* of Keynesian-welfarist and social-collectivist institutions", "roll-out" neoliberalism is defined by "the purposeful *construction and consolidation* of neoliberalized state forms, modes of governance, and regulatory relations" (Peck and Tickell, 2002: 384, emphasis in original). The strategic deployment of "community" is one among several "extramarket forms of governance" (Peck and Tickell, 2002: 390. See endnote 16).

The process of neoliberalization, in turn, is supported and indeed propelled by a discursive formation that produces neoliberalism and the market as the most efficient, or "*best*" (Castree, 2008a: 143, emphasis his), means for creating and distributing wealth globally. What critics regard as a deeply political project is thereby recast in the politically neutral and normalizing language of economic technicality, of obviousness, of common sense. "[T]he hegemony of neoliberalism", write James McCarthy and Scott Prudham (2004: 276), "is made most evident by the ways in which profoundly political and ideological projects have successfully masqueraded as a set of objective, natural, and technocratic truisms". However, political action against neoliberalism in its diverse forms "gives the lie to such disguises", they continue, "exposing the political negotiations and myriad contradictions, tensions and failures of neoliberalizations" (McCarthy and Prudham, 2004: 276).

I return to the question of resistance shortly but, in order to situate community land ownership and debates about property and nature more precisely with reference to processes of neoliberalization, I focus initially on enclosure and the privatization of property rights, before considering the ways in which the move from common rights to individual or private rights to property is at the same time about the reconfiguration of nature/society relations.

Privatization, which follows processes of enclosure, writes Becky Mansfield (2007a: 396), "is not merely one of several shifts promoted under neoliberalism, but instead is the central assumption and precursor to other market-based reforms". It is "a precondition for capitalist commodification" (Castree, 2003: 279). It refers to

> a process through which activities, resources, and the like, which had not been formally privately owned, managed or organized, are taken away from whoever or whatever owned them before and transferred to a new property configuration that is based on some form of "private" ownership or control (Swyngedouw, 2005: 82).

This mode of expanding capital – "primitive accumulation", or what David Harvey (2003) calls "accumulation by dispossession" – is, Erik Swyngedouw (2005: 82) argues, "nothing less than a legally and institutionally condoned, if not encouraged, form of theft". It "equates [with] a process of 'dispossession'", he continues, impelled by discursive formations that "[render] such acts of theft not only legitimate, but normatively desirable" (Swyngedouw, 2005: 82). On the one hand, these mobilize ideas of "state failure" or, more generally, the "'failure' of non-private modes of social organization of production"; on the other, they call on such "moralistic arguments" as those of Garrett Hardin's (1968) "The tragedy of the commons" in order to justify a move towards private ownership (Swyngedouw, 2005: 82–83).[14]

The process of dispossession proceeds on the basis of a model of property that recognizes, primarily, private rights of ownership, viewing these rights as exclusive, absolute and alienable, the state having the right to "intervene" in the exercise of these rights only where they "threaten harm to others" (Blomley, 2004: 2). Through this construction of property rights in the either/or terms of private/public, other rights that do not involve ownership – of use, for instance – or that are held in common become "invisible", "unintelligible" (Mackenzie, 1998a) or, in

James Scott's (1998: 37) words, "illegible" to the state. Where they do appear, it is only to be dismissed as "dysfunctional" or "anomalous" (Blomley, 2008: 317). The research by Tom Flanagan and Christopher Alcantara (2004) into property rights on First Nations' "reserves" in Canada is a recent case in point. "Markets work best", these authors write (2004: 529), "when property is privately owned". Thus, their study concludes, collectively owned land must give way to private ownership if the land if is to yield its "maximum benefit" to First Nations peoples (2004: 530).

This view of property – called by Joseph Singer (2000) "the ownership model" – draws on an intellectual trajectory traceable to John Locke's treatises of the late seventeenth century and his conjecture that processes of enclosure which lead to private ownership are "not simply inevitable, given the unfolding telos of property [but they are] also normatively *good*, to the degree that [they express] divine will" (Blomley, 2004: 85–86, emphasis in original). For Locke, "individual, private property rights guaranteed by the state constituted the foundation of a just and efficient social order", "a society better for all" (McCarthy and Prudham, 2004: 277). It was this construction of political economic order, together with the accompanying normalizing discourses of "betterment" or what has more recently been cast as "development" and "progress", that underpinned the imperial project (Cowen and Shenton, 1996; also, see Mudimbe, 1988; Escobar, 1995; Mackenzie, 1998a). It is "a vision", McCarthy and Prudham (2004: 277) argue, entirely compatible with neoliberal claims, whether these concern the land or water, genetic resources, environmental management (tradable carbon credits and fishing quotas are examples) or local knowledges.

As evident in the violences of colonial rule as well as those perpetrated in "the colonial present" (to extend the use of Derek Gregory's, 2004, troubling phrase), the ownership model of property "polices" the boundaries of what counts as property and who counts as owner (Blomley, 2008: 321). Where private ownership does not prevail, the existence of any (collective) rights to property may be refused, on grounds that are far from innocent. What Nicholas Blomley (2002) refers to as "the master narrative of property" erases or renders unseeable other claims and, through the imaginary construction of a *terra nullius*, acts as a political technology, legitimating the dispossession of peoples with long-standing rights to particular territories and other "resources". That this act of dispossession may be gendered as well as being constituted through constructions of race and class is evident (see, for example, Mackenzie, 1998a).

A counterhegemonic narrative of property requires destabilizing the ownership model by disrupting the norms – the binary private/public – through which it is constituted and by tracing the ways through which people thereby reposition themselves, as subjects, towards property. Destabilizing this model involves two interwoven processes. First, it is necessary to make visible those processes that were silenced through the "master narrative". It involves recognizing that there are other rights than private rights to property, that these rights might or might not involve ownership and that there might well be overlapping rights to any particular territory. Centrally, it involves counterposing "the commons" or, more precisely, rights held in common or collectively, to the binary. Following Blomley (2008: 318), common property is used here to refer to a situation where "a resource is held by an identifiable community of interdependent users, who exclude outsiders while regulating internal use by community members". Defined thus, common property is distinguished from "open access" regimes, which are not subject to any form of ownership or control (Mansfield, 2007b: 67), and which were the subject of Hardin's (1968) polemic.

While many forms of commons exist – within as well as outwith state law (Blomley, 2008: 317), some politically progressive, others regressive (McCarthy, 2005a: 18–19) – the common property regime literature has tended to focus analytically on the economic and institutional dimensions of common property (for example, Ostrom, 1990; Ostrom *et al.*, 1999), informed, as Blomley (2008: 318) points out, by rational choice theory. One result of this "economic logic", he shows, is that "crucial political and ethical dimensions" of the commons are ignored (Blomley, 2008: 318) and political possibilities compromised. To illustrate the point, Blomley (2008) refers to the struggle over rights to the site of Woodward's, a store in Vancouver's Downtown Eastside (British Columbia, Canada), closed in 1993 and subsequently caught up in plans for "redevelopment" and gentrification in the mid-1990s. The interests of private developers were pitted against Downtown Eastside's community of long-standing, (predominantly) low-income residents, who feared displacement with a rise in property values. Countering the private developer's claim of a right "to exclude" (Blomley, 2008: 316), Blomley (2008: 318) shows how residents drew on "a moral and political commons, justified and enacted through a language of rights and justice", not "an instrumental commons, governed by rules", to assert their claim to the site. Such commons, he continues, are "sustained by deeply entrenched values and beliefs" (Blomley, 2008: 319). In the case of Vancouver's Downtown Eastside, Blomley (2008: 320) proposes, the claim of the poor "is based upon and enacted through sustained patterns of local use and collective habitation, through ingrained practices of appropriation and 'investment'". Recognizing this commons, he continues, "is a crucial political task through which non-capitalist possibilities can be discerned and revalorized"; it "prise[s] open" a "space of hope and potentiality" (Blomley, 2008: 322). In addition, locating particular claims, such as that of Woodward's, within the wider "global commoners' movement", he adds, is "ethically useful" (Blomley, 2008: 324). A politics of a local commons is not necessarily progressive: it may support a neoliberal agenda – a point also made by McCarthy (2005a: 18–19), whereas "sutur[ing]" local claims to a commons to a global movement protesting against a neoliberal form of globalization can work against this (Blomley, 2008: 324, drawing on Klein, 2001).

Karen Bakker's (2007a) research into water provision in the global south provides a further example of how this more complicated view of the commons plays itself out politically in the struggle against privatization. She shows how activists disturb the public/private binary through which struggles over water provision are customarily waged by countering the state's and corporate interests in water as a commodity with "a commons view" of water as "a flow resource" essential for the mutual well-being of people and ecosystems bound together, collectively, through the hydrological cycle, as "non-substitutable", having cultural meaning and spiritual value (Bakker, 2007a: 441). Activists, she recalls, position their claim against the conscription of water as a commodity in both state and corporate discourse as a common property right rather than as a human right, as "'rights talk' resuscitates a public/private binary" producing only two unsatisfactory alternatives – "state or market control" (Bakker, 2007a: 440). Human rights, she notes, are "individualistic, anthropocentric, state-centric, and compatible with private sector provision of water supply; and as such, a limited strategy for those seeking to refute water privatization" (Bakker, 2007a: 447). It is rather the commons that is the appropriate "antonym" in the struggle against the commodification of water, not human rights, and it is a commons-based claim that thereby contributes to the production of a place-based "alter-globalization" movement (Bakker, 2007a: 436).[15,16]

Destabilizing the hegemony of the ownership model of property, as shown by the research I have cited, involves a search for places where old commons (water) are recaptured or new ones (Woodward's) conjured. But, second, it requires undoing the terms of the binary through which it is produced. Private property is presented in the ownership

model as an individual or corporate right – absolute, or allodial, exclusive
and alienable – to a "thing", a move that, as Blomley (2005: 126)
explains, allows property to be depoliticized. But, as he and others point
out, property is very much bound up with social relations and, at times,
political struggle. It is, Blomley (2004: 13) writes, "simply inaccurate" to
suggest that an "owner" is able to exercise "absolute control" over his or
her property. There may be "mortgage lenders, neighbors (who may
have rights recognized by law), spouses and those granted easements"
who also have particular rights to a property. In the postcolonial city, as
well as in countless non-urban territories, "ownership" rights may be
caught up in struggles over aboriginal title and who can claim the right
of sovereignty. There may, in other words, be competing and overlapping
claims to property. Whether in urban or rural spaces, not only are notions
of "private rights" and outright "ownership" profoundly disturbed, but
what counts as "the state" and who has jurisdictional authority over
land are called into question.

The distinction between private rights and the state is also less precise
than that presumed in the ownership model. Blomley (2004: 13) gives
numerous examples, some sanctioned by the state, others not. One
example of the former concerns zoning laws, such as those that involve
building codes or place restrictions on tree removal. However, also
included here are examples where collective claims to property are
made, as discussed above. Where these trouble the right to alienate – i.e.
where property is removed as a commodity from circuits of local or
global capital, as in the case of community land trusts (see Abromowitz,
2000) – there is, of course, a direct assault on processes of enclosure and
privatization. Squatter settlements provide a further example of the
muddying of the boundaries between the private and the public. Their
complexity and ambiguity *vis à vis* the ownership model is illustrated by
a case in the UK. "Normally" cast as operating outwith state sanction,
Blomley (2004: 21) indicates how an "inversion" occurs when the UK
Advisory Service for Squatters provides squatters with "a template 'legal
warning'", which cites the Criminal Law Act (1977) and which can be
used by them to assert rights to their squat. As a final example of
troubling the boundaries between public and private rights, and one
that I will discuss further in Chapter 2, under feudal law – as pertained
until very recently in almost all privately owned land in Scotland – there
is no "separation", "at a technical level", between public and private
rights, because of the rights retained by the Crown (Callander, 1998:
33). The owners of the land, writes Robin Callander (1998: 9), "do not
own their land outright and their authority over that land has always

been constrained not only by the general laws of the country, but also by their feudal title to their land". The result is that private rights are always "inherently conditional on the public interest" (Callander, 1998: 33), a characteristic that both Callander and sociologist David McCrone (1997) argue provided openings for rethinking land rights in future land reform.

In this discussion, the idea of property is rendered more "capacious" and "heterogenous" than is the case in the ownership model (Blomley, 2004: 15). The binary public/private, which is produced as stable and predetermined, common sense and natural, in that model, is called into question. Whether "private", "state" or "common", property is shown to be always and actively "doing" (Blomley, 2003: 122), its enactment bound up in the complications and complexities of social relations. It is, in other words, deeply political. It may be about "theft"/privatization and commodification; it may also, Blomley (2004: 15) argues, contain within it "radical potential", moments of political possibility when the dominant norms are interrupted and ongoing practices of enclosure reversed or at least troubled. What is then needed in analytical terms is to keep open to continuous resignification the complex, frequently contested and constantly changing meanings of property as land is brought into community ownership in order to explore the political possibilities of the commoning of the land.

In this book, I take these ideas about the "doing" of property to trace the ways through which bringing land into community ownership "undoes" dominant norms of property that rely on practices of privatization and enclosure and on land as a commodity with exchange value in circuits of global capital (Chapter 2). To extend Gibson-Graham's (2006) theorization of a diverse economy, rethinking property in this way opens up the political possibilities for people to reconfigure their individual and collective subject positions in the search for more just futures. It contributes to the re-creation of the "commonweal", a Scots word meaning the common good or common well-being.[17]

This disturbance of the neoliberal norms through which property is so frequently produced is not only about reversing processes of enclosure that began hundreds of years ago. It is also about the creation of material and metaphorical space through which people's relationship with "nature" can be reworked. Privatization is about property, as I have shown, but it is additionally about the reworking of the relations between people and nature such that the process of capital accumulation is deepened. It is through privatization – of property – that, to borrow Cindi Katz's (1998: 46) phrase, nature becomes "an accumulation strategy for capital". Here, I identify first, briefly, the ways through which nature is produced through neoliberalization in the interest of capital accumulation before, second, indicating how troubling property opens nature to new, more socially just, significations.

"[N]ature's neoliberalisation", suggests Castree (2008a: 150, emphasis his), proceeds in ways that appear paradoxical: it is at once "about conservation *and* its two antitheses of destroying existing and creating new biophysical resources". To consider, first, conservation, nature is conscripted as part of a neoliberal agenda in two ways. The first, referred to as "free market environmentalism" by its supporters, involves the privatization and marketization of environmental management (Castree, 2008a: 147). Examples include wetland mitigation banking in the USA (Robertson, 2007), the provision of water supplies in England and Wales (Bakker, 2007b), and the conservation of fish stocks by the USA through Individual Transferable Quotas in the case of the North Pacific (Mansfield, 2007b) or, in the case of New England fisheries, through a series of measures that include restrictions on licences for specific species, on type of gear permitted, on access to fishing grounds and on the number of allowable days at sea (St. Martin, 2007).

The second way through which conservation reworks nature concerns the growing number and extent of parks or biosphere reserves, particularly in the global south. "Debt-for nature" swaps, where the debt of an impoverished country in the global south is exchanged, at a discounted rate, by states or non-governmental organizations in the global north for the "preservation" of a particular area, provide another example (Katz, 1998: 50). The "intent" of these areas of conservation, to paraphrase Katz (1998: 47), is to effect the enclosure of specific territories in ways such that there is the simultaneous erasure of specific histories and the production of a space for "bio-accumulation". She writes:

> Underwriting these strategies are deeply problematic constructions of nature that turn around peculiar and problematic tropes of wild and wilderness; a class-based, racialized, and imperially inflected notion of the "public" and its "commons"; and a paradoxical understanding of material social practices as somehow outside nature (1998: 48).

For Arturo Escobar (1996), such conservation strategies work, discursively as well as materially, as a form of "postmodern ecological capital". Distinguished from the "modern [i.e. exploitative] form of ecological capital", which proceeds through such normalizing discourses as those of science and progress, the postmodern form, advanced by discourses of sustainability and biodiversity, opens up new territories, new communities and local knowledges to capital (Escobar, 1996: 56–57). People and communities, argues Escobar (1996: 57), may be valued as "stewards of nature" or as the repositories of valuable knowledge (for example, for bio-prospecting by pharmaceutical companies), but the cultural complexity and embeddedness of this knowledge – including its frequently gendered and generational dimensions – are ignored. It is coded in ways that are economically reductionist.

The neoliberalization of nature also proceeds through the exploitation, or degradation, of biophysical resources – Escobar's modern form of ecological capital. Through ongoing processes of enclosure and privatization of land, forests, oil, fish stocks and, for the purposes of producing energy from renewable sources, the wind and water from rivers and the sea, nature is recruited in the interest of capital accumulation. Exposing more of the "nonhuman world" to the market, as Castree (2008a: 147) notes, "overlaps closely" with the strategy of "accumulation by dispossession" (Harvey 2003) identified earlier. Examples based on detailed case study research abound and include contributions to special issues of *Geoforum* (2004, 35(3)), *Capitalism Nature Socialism* (2005, 16(1)) and *Antipode* (2007, 39(3)), some of which have been re-published in modified form in the book *Neoliberal Environments* (2007a), edited by Nik Heynen, James McCarthy, Scott Prudham and Paul Robbins.[18]

The creation of new biophysical resources provides a final way through which nature may be enlisted in the process of neoliberalization. Genetic engineering is one obvious instance of the extension of property rights to new forms of nature. Sarah Whatmore's (2002) forensic examination of the soy bean details the precise ways through which nature is reinvented in the interest of capital. Genetically modified seeds, as is evident from her analysis as well as from globally diffuse protest that has greeted their field trials and distribution, are firmly inscribed in political process – in dispossession. Parallel arguments can be made through Scott Prudham's (2007) work on patents in the cases of Harvard College's oncomouse and Monsanto Corporation's "Roundup Ready" canola heard by the Supreme Court of Canada in 2002 and 2004 respectively.

Critical scholarship such as that identified above disputes the idea, still prevalent within as well as outwith academia, that nature is ontologically separate from, or external to, society. Projecting the origins of this dualism back to Francis Bacon in early seventeenth century England, Neil Smith (2008: 11) refers to this conceptualization of nature as "a thing, the realm of extra human objects and processes existing outside society. ... [It] is pristine, God-given, autonomous; it is the raw material from which society is built, the frontier which industrial capitalism continually pushes back". It is this construction of nature, as Katz (1998: 48) identifies in the quotation cited earlier concerning conservation practices, that underpins processes of enclosure and privatization. As its counter, captured through the phrase "social nature" to which I turn shortly, it comes with political baggage. Claiming legitimacy on the grounds of "scientific method" with its normalizing procedures of "measurability", "objectivity" and "replicability", a politics of nature as external allows the pursuit of particular interests while at the same time silencing others. In ways that parallel the operation of the ownership model of property, a politics pursued through the dualism nature/society hides the ways through which power operates. It acts as a political technology (Foucault, 1979), casting as asocial and apolitical a relationship between society and nature that, its critics insist, is deeply and interrelatedly social and political – with respect to class, race, gender, among other axes of social differentiation. Dangerously so with respect to the practice of "wilderness", or so William Cronon (1995: 80–81, emphasis his) argues in his influential essay on the subject, the dualism produces "wilderness" as the place where people are not, as the only place where "true" nature is found, leaving us with "little hope of discovering what an ethical, sustainable, *honorable* human place in nature might actually look like". He continues:

> In its flight from history, in its siren song of escape, in its reproduction of the dangerous dualism that sets human beings outside of nature – in all these ways, wilderness poses a serious threat to responsible environmentalism at the end of the twentieth century (Cronon, 1995: 81).

"The costs of retaining the dualism", write Castree and Braun (1998: 34), have simply "become too high; as Latour explains, too much is left unseen".

Theorists of social nature who, like Castree and Braun, recognize that, at every turn, the social is inevitably and integrally intertwined with the natural, seek both to expose the political and social practices through which nature is constructed and to suggest what a different politics to that based on the binary society/nature might look like. Researchers such as those identified in the earlier discussion of the neoliberalization of nature draw variously on a theoretical repertoire that includes insights from poststructural critiques of Marxian theorizing of the production of nature, from Butler's work on performativity and (following Foucault) the productivity of power, and from the field of the sociology of scientific knowledge, where writers such as Bruno Latour and Donna Haraway interrogate the "hybridity" of social natures (Castree and Braun, 1998: 15–33). Although the political possibilities that may emerge from these richly diverse investigations are frequently not traced to any substantial degree, as Castree (2008b) has noted, it is evident that the concept of social nature provides both "analytic and political hope" (Braun, 2002: 10) in a context of the ongoing processes of neoliberal globalization. Analytically, this approach requires that attention be given to "the *specific historical forms* that nature's production takes" and "the specific *generative forms* that shape how this occurs" (Braun, 2002: 11, emphasis his). In terms of politics, working with the idea of social nature, argues Bruce Braun (2002: 13, emphasis his), "forces us to take responsibility for *how* this remaking of nature occurs, in *whose* interests, and with *what* consequences (for people, plants, and animals alike). It brings together ecology and social justice".

As a reminder to those environmentalists who are sceptical, Braun (2002: 14) insists that this is not a matter of condoning "all human environmental practices". "There is no reason", he argues, why an environmental movement informed by social nature "should be any less concerned with the health of the planet and its many inhabitants, human and nonhuman alike, than any other ecopolitics" (Braun, 2002: 14). Instead, such a politics may lead to "a reinvigorated environmentalism":

> [Social nature] brings society and ecology together into a single analytic field, allows us to critically examine and evaluate the many ways that nature is socially produced, and draws attention to the ways in which nature's production – including its preservation – is always entangled with much more than nature, including questions of class, race, gender, and sexuality. It does not dictate to us what future natures *should* look like, nor does it provide a template for developing normative statements about nature and its transformation; these are open-ended questions that will be decided by the play of historical forces and political struggle (Braun, 2002: 14).

This different politics is not then about "policing boundaries" between an ontologically separate "nature" and "culture" (Castree and Braun, 1998: 34). It is about disturbing the norm of nature as external to the social and opening up the meanings of the terms "nature" and "the social" to allow new ways of thinking critically and creatively about how to move forward, about "what kinds of marks we wish to leave", to borrow Cronon's (1995: 88) phrase.

I employ the concept of social nature as a way of exploring the ways through which nature is reworked once the land is brought into community ownership. I am interested in the political possibilities of a nature that is now produced through a reversal of practices of enclosure and privatization of the land. There are two parts to the argument. First, I am concerned to trace the extent to which a disturbance of the norms captured in the binary private/public with respect to property rights troubles the dualism nature/culture in so far as it concerns practices of conservation (Chapter 3). As I show later, substantial areas in the Outer Hebrides and specifically the Isle of Harris are subject to protective environmental designations whose provenance lies elsewhere – at national (Scotland), UK and European levels. To what extent do new practices of nature by land owning community trusts – that have to do with these areas of conservation – write against "the scientific colonialism" of "'core' conservationists" (Mather, 1993: 374), underpinned as this approach to the preservation of "wild" places is by the dualism nature/culture? Examining such recent practices as the planting of "native" woodland, the management of environmentally protected areas, and the proposal for a national park on Harris, I consider how a community land owning trust "undoes" a norm of nature produced through this dualism. I demonstrate how the land owning trust extracts the idea of the "wild" from colonizing – and class-based – configurations of conservation and opens up the meanings of nature and the wild to new configurations, thereby creating the space for the renegotiation of a community subjectivity.

Second, I draw on social nature theorizing to investigate the process of commodification of the wind, now a highly valued resource both in the effort to turn around local economic (mis)fortunes in the Outer Hebrides and in the attempt to achieve renewable energy targets within Scotland and, more generally, the UK and its commitments to international protocols (Chapter 4). Through plans to build wind farms and thus capture the wind, nature enters the political arena in complex ways that centre on the question of land ownership. Who owns the land has the right to work the wind, as I have discussed elsewhere (Mackenzie, 2006b). Unlike many other struggles over a commons, it is not simply a case of (local) resistance to a process of enclosure – the commodification of a resource – although there is indeed substantial opposition to attempts by corporations and private individuals (land owners) to erect large-scale wind farms. Such attempts may unequivocally be analysed as a process of "accumulation by dispossession" (Harvey, 2003), legitimated through the planning process on account of a gesture towards a local community. However, where the surplus from the production of electricity and its sale to the national grid is distributed according to the priorities of a community land owning trust, rather than to distant shareholders or to an individual owner's bank account, the norms of commodification are disrupted. Here, aligned with a collective rather than corporate ethic or individual interest, the process of commodification of the wind reinforces collective rights to the land rather than undercutting them. It is at this point of troubling the norm, I argue, that the political possibilities of a new social nature emerge.

In Chapter 5, I extend the argument about the political possibilities created when norms of property and nature are disrupted through community land ownership by examining a series of initiatives that work together to re-create community and place. I show how the activities of the North Harris Trust – by building houses, reducing the community's carbon footprint, promoting local food production, conducting archaeological research, and restoring a network of footpaths - produce a new way of seeing the estate. The activities, as those I have discussed in previous chapters, conjure an optic and a counternarrative that call into question the hegemony of processes of neoliberalization. They suggest the exercise of a right that, to borrow from Castree (2004: 136), allows people, collectively, to "make their own places, rather than have them made for them". By reworking the meanings of a land now held in common, they point towards a more radical, socially just, and sustainable engagement with the future than was possible before community land ownership.

Notes

1   The figure of about 425 000 acres in community ownership is more than the acreage owned by the National Trust for Scotland, the Royal Society for the Protection of Birds in the Highlands and Islands, the John Muir Trust and the Scottish Wildlife Trust combined (Cameron, 2009: 15).

2   I am excluding here the Stornoway Trust set up in 1923 after the owner of Lewis and Harris, William Hesketh Lever, Lord Leverhulme, offered the estate to the Stornoway Town Council (see Hutchinson, 2003). The Assynt Crofters' Trust is generally considered to be the reference point for the contemporary community land movement in Scotland. For a discussion of the origins of the Stornoway Trust – at a time of land struggle

within Lewis and of the fiscal crisis Leverhulme faced – see the work of Joni Buchanan (1996) and Roger Hutchinson (2003). Its detailed story remains to be told. The Stornoway Trust is now a member of Community Land Scotland (Chapter 6).

3 According to Wightman (2010: 106–107), of the 15 722 287 acres of rural land in private ownership in Scotland (out of a total rural acreage of 18 924 516), 9.4 million acres are owned "by a mere 969 landowners" and over 10 million acres are held by 1550 private land owners in estates of 1000 acres or more. With 83.1 per cent, the private sector is by far the largest land owner. The public sector owns 12.1 per cent of rural land, the heritage sector 2.5 per cent and the community sector 2.2 per cent (Wightman, 2010: 106).

4 The play was staged by the 7:84 Theatre Company, whose name refers to the percentage of the UK population, 7 per cent, that owns 84 per cent of the wealth.

5 A croft is often humorously referred to as "a small area of land surrounded by regulations" (a crofter's son, cited by J. MacDonald in *A Short History of Crofting* (1998) in Busby and Macleod, 2010: 602). As a tenant of an estate under crofting tenure, a crofter has use rights to inbye land, ranging in extent from under half a hectare to over 50 ha, but averaging about 5 ha (Crofters Commission, 2011) and rights to common grazings, which may extend to thousands of hectares. These grazing rights are shared with other members of the crofting township. (For a definition of a crofting township, as given in the Land Reform (Scotland) Act 2003, see Chapter 2.) Security of tenure, the right of succession, the right to the value of any improvements carried out on the croft and the right to a fair rent have, among other measures, been assured since the Crofters' Holdings (Scotland) Act 1886. There are at present 17 923 crofts in Scotland supporting a population of about 33 000 (Crofters Commission, 2011). Crofts extend over about 17 per cent of Scotland's land base and are located in what are generally referred to as "the crofting counties": Shetland, Orkney, Caithness, Sutherland, Ross-shire, Inverness-shire and Argyll (Committee of Inquiry on Crofting, 2008: 16). The Crofting Reform (Scotland) Act 2007 expanded the area under the jurisdiction of crofting law to encompass the entire Highland Council area, Moray, the parishes of Kingarth, North Bute and Rothesay in Argyll and Bute, and the islands of Arran (including Holy Island and Pladda), Great Cumbrae and Little Cumbrae in North Ayrshire (Macleod *et al.*, 2010: 97). By extending the area where new crofts and common grazings could be designated, the 2007 legislation also extended the potential reach of Part 3 of the Land Reform (Scotland) Act 2003 (see Chapter 2).

6 First, the delegates proposed that public funding be once again made available for land purchase by communities. They called on the Scottish Government to re-establish the Scottish Land Fund that had been so key to the earlier community land buyouts and to finance it through taxation

revenue. Second, they proposed that the Community Land Unit of Highlands and Islands Enterprise – so central in providing financial and logistical support for community land purchase – be given the necessary resources by Scottish ministers once again to take up this task (for an evaluation of the Community Land Unit, see SQW, 2005). And third, recognizing the particular difficulties faced by communities wishing to purchase land owned by the government, the delegates proposed that ways be found to facilitate this process at no, or minimal, cost.

7 The Council retains the title of the Western Isles or, more usually, its Gaelic equivalent, *Comhairle nan Eilean Siar*, but with the current "re-branding" of the islands they are commonly referred to as the Outer Hebrides or, in Gaelic, *Innse Gall* (The Isles of Strangers, so named after the period of Norse settlement) (MacAulay, 1996: 6–7, cited by McIntosh, 2004: 18). The earlier name, *Innis Bhrighde* (Isles of St. Bridgit), John MacAulay (1996: 6–7) notes, disappeared a long time ago.

8 These figures compare with an average of 61.1 per cent for the Outer Hebrides, "the Gaelic heartland of Scotland", as a whole (Bryden *et al.*, 2008, Appendix 5: 21, 23).

9 It is necessary, of course, to recognize that Scotland's Highlands and Islands were not colonized in the way that, for example, many parts of Sub-Saharan Africa were (see Hunter, 1995a: 28). It is nevertheless the case that the same discourses – of "progress" and "development", as examples – that propelled the imperial project in places farther afield were deployed with devastating material effect in the Highlands and Islands, at times directed from London, at others, from Edinburgh (Hunter, 1995a: 28).

10 Foucault's (1990) focus here is on "morality". He distinguishes between the meaning of morality as "a code", "a set of values and rules of action" and "the real behavior of individuals in relation to the rules and values" (1990: 25). With reference to the second, he writes that, "There is no forming of oneself as an ethical subject without 'modes of subjectivation' and an 'ascetics' or 'practices of the self' that support them" (1990: 28). Modes of subjection he defines as "the way in which the individual establishes his [sic] relation to the rule and recognizes himself [sic] as obliged to put it into practice" (1990: 27).

11 In order to capture more precisely the ongoing process of resubjectivation, I have replaced Nancy's (1991: 4, emphasis his) words, "being *in* common", with the phrase, "becoming in common". Gibson-Graham (2006: 85, emphasis hers) notes that for Nancy there was no "common being", as that phrase recalls a community identity that is "already known [which] precludes the *becoming* of new and as-yet unthought ways of being".

12 Massey (2009: 415) makes it clear that such a politics is not "necessarily progressive" – as evident in Bush's and Blair's war in Iraq.

13 For an excellent introduction to poststructural political ecology, see Peet and Watts (2004).

14 These "moralistic arguments" frequently confuse, as did Hardin, "open access" situations with common property regimes. It is the former, rather than the latter, where the "free rider" problem emerges (see, for example, Ostrom, 2005: 80).

15 The "most progressive strategies" for the provision of water, Bakker (2007a: 446) suggests, are those that involve both government reform and the building of local, community-led, institutions for resource management.

16 In the effort to deploy the commons politically in the struggle against privatization, Bakker (2007a: 443–444) is concerned to distance herself from a "romantic" view of "community" – an entity that is socially and economically homogeneous, beyond the fray of politics and always successful in sustainably managing resources. She points both to literature that supports claims of effective community management of such a resource as water and to research that documents its shortcomings (Bakker, 2007a: 446). In addition, it is also necessary to recognize how frequently "community" is now conscripted in neoliberalizing discourse. In reference to Wendy Wolford's (2007) research into the land struggle in Brazil, Michael Watts (2007: 277) notes that "the invocation of 'community' as a counter-force to market-led reform fails to come to terms with the extent to which the community has become the neoliberal form *par excellence* of modern governmentality". The critical issue to explore is whether calls for "community" and a new "commons" are indeed part of a counterhegemonic imaginary to that of enclosure and privatization or whether they are thereby coopted into the prevailing neoliberal imaginary (see McCarthy, 2005a: 18–19), whether they contribute to the creation of a more diverse, less capitalocentric economy as suggested by Gibson-Graham (2006), or to the furthering of neoliberal norms. There can be no "global, universal evaluation" of this issue, writes McCarthy (2005a: 19).

17 The term has recently gained increased political visibility through The Commonweal Project of the Caledonia Centre for Social Development. The project aims to document and disseminate knowledge about common property rights in Scotland, and thereby contribute to the political case for asserting common property rights, particularly for those whose well-being is at risk with their loss (see, e.g., Reid, 2003).

18 For a critical discussion and synthesis of some of this published work, see Castree, 2008a, b.

2

Working Property

"'S Leinn Fhèin am Fearann" ("The land is ours"): a day to celebrate

High above Loch Shiphoirt (Seaforth) on 21 March 2003, a blustery but dry day, some 200 people gathered on the hillside at Aird a'Mhulaidh to celebrate the handing over of the North Harris Estate to the community of North Harris. Led to the site by two pipers, daughters of Simon Fraser, the "legal architect" of crofting community land buyouts in the Highlands and Islands (Currie, 2003: 7), the ceremony began with the welcoming address of the Chair of the North Harris Trust, Calum MacKay.[1] With the purchase of the estate, the North Harris Trust, he said, "was making history in a number of ways". First, the land had been bought by a community trust, not a private land owner as had been the previous norm. Second, accounting for 55 000 acres, this was the largest community buyout in Scotland to date. And third, the agreement to purchase had also involved a legally separate "partner". Ian Scarr Hall's decision to take on the ownership of Amhuinnsuidhe Castle, together with 600 acres of adjacent land and the salmon fishings, had made the community's bid for the land *without* the castle possible (Mackenzie, 2006a).[2]

In a brief ceremony to mark the transfer of the land, Simon Fraser invoked a legal practice that, he claimed, had existed "from time immemorial" in Scotland. It was "the custom and the law", he said, to

Places of Possibility: Property, Nature and Community Land Ownership,
First Edition. A. Fiona D. Mackenzie.
© 2013 A. Fiona D. Mackenzie. Published 2013 by Blackwell Publishing Ltd.

Photograph 2.1   Recording the deeds, 21 March 2003.
Source: The North Harris Trust. Used with kind permission.

hand over stone and earth in the presence of witnesses on the land itself
to mark the passing of the land to the new owners. The ceremony, known
as "giving sasine", he recounted, had been abolished as a written record
of the transfer gained in importance but, as in Gigha the previous year,
it was now to be re-enacted[3] (Photograph 2.1). Simon Fraser called on
David Cameron, former Chair, to receive sasine on behalf of the North
Harris Trust:

> I hereby deliver into your hands, stone and earth of this land and in doing
> so give unto the North Harris Trust true and lawful sasine of these whole
> Lands of North Harris; from the low tide of the sea to the highest mountain
> tops, *ab caelo usque ad centrum* [from the sky to the centre of the earth];
> to be held on behalf of the people of North Harris in all time coming; *agus
> tha sinn uile a tha cruinn còmhla an seo an dràsda a' guidhe gu soirbhich
> leibh agus leis gach ginealach a tha ri teachd agus gum faigh sibh a h-uile
> beannachd anns an talamh seo a tha ar tighearna air a thoirt dhuibh ri
> shealbhachadh* (Currie, 2003: 7).[4]

Following the planting of a tree – a rowan – by Joanna Morrison, a young woman from a crofting family at Cliasmol, assisted by David Cameron, people migrated to the shelter and warmth of the nearby Scaladale Outdoor Centre to continue the celebrations. Of a number of speeches – from politicians, from the Scottish Land Fund, the Comhairle nan Eilean Siar/The Western Isles Council, the Highland Council/ Comhairle na Gàidhealtachd, the John Muir Trust – I draw here on two, one by a local crofter, the other by a member of parliament (Westminster), to demonstrate something of the historical significance of the event and to set the stage for the remainder of the chapter. I am interested in exploring "unruly pasts" – metaphors and material practices through which the land has been conjured in the effort to maintain a collective or common right.

Murdo Morrison, a crofter from Aird a'Mhulaidh, was the first to speak at the Scaladale Centre. "It is hard to imagine", he recalled, "that 116 years ago government gunboats came into Loch Seaforth to deal with land raiders in Pàirc", the estate on the opposite side of the loch to the Centre, on the Isle of Lewis. HMS Seahorse and HMS Jackal, each carrying 40 marines, and HMS Ajax with 400 marines on board, had been dispatched to quell an "uprising" of seven or eight impoverished cottars who had raided the land (a "deer forest") owned by Lady Matheson.[5] Such a disproportionate threat of state force – the incident ended peacefully – was the cause of mirth as people applauded Murdo's story of the raiders' bravery. These events in Pàirc, Murdo Morrison continued, were witnessed by two young men, Angus and John Morrison, his grandfather and great uncle, respectively. In 1916 and 1917, they too raided land, in Aird a'Mhulaidh – working it continuously until their right to use the land was recognized legally in 1922. Forty years later, consistent with other tenants of crofts, they were finally given rights to common grazings, on part of which land the Scaladale Centre now stands. With respect to the land raids in Pàirc – and however open to dispute are the numbers of the raiders and marines – the story is a sobering reminder of the violences that can accompany the exercise of rights to private property.[6] In the cases of both Pàirc and Aird a'Mhulaidh, people's resistance to what they perceived as dispossession of land to which they had rights in common is made abundantly clear.

Brian Wilson, then Energy Minister in the Westminster Government and a lead figure in the land reform movement in Scotland, put the events of which Murdo Morrison spoke and the day's celebrations in the broader historical context. It was "a day which will have a special place in the history of the Gael", he began. He continued,

It is not strictly true, I have to say, that this is a dream come true with community ownership of the North Harris Estate because, until recently, it was a prospect which did not even feature in our dreams and aspirations. North Harris, by virtue of its scale and history, was the epitome of private land ownership in the Hebrides, a fortress that would never crumble. And yet, on this great day, ownership and all the potential that goes with it are passing into the hands of the North Harris community. If anyone doubts that the trickle of land reform in the Highlands is going to become a tide, then they should observe what has happened in North Harris over the past few months. Ownership and management of these estates, by the people who live on them, is a noble concept whose time has come.

Wilson concluded by recounting a story of an earlier landlord of the North Harris Estate, Sir Hereward Wake. Thirty years ago, there was "great controversy" over what he called "a surreal proposition", namely Wake's plan to build a bypass around Amhuinnsuidhe Castle. The road passed, as it still does, directly in front of the castle, thus disturbing his view, and privacy. "By fruitful coincidence", Wilson recalled, both Wake and the Chair of Inverness County Council's Roads Committee, Lord Burton, had been at Eton together and both were members of the highly elitist Brooks's Club in London. Citing the words of Derek Cooper's poem on the subject, Wilson recalled, how money – "forty thou'" (£40 000) – passed hands "to see things through committee". The ceremony ended with a short service of thanksgiving and the singing, in Gaelic, of Psalm 24.

Unruly properties

However rigorously the North Harris Trust and other community land owning trusts had been required to justify their claim to the land on the grounds of economic feasibility in order to secure public funding for the purchase, there was no mistaking the sense of history evident in the opening ceremonies. Above, I have identified glimpses into both a history of dispossession and of resistance to it at times when private rights to land prevailed over those of a collective interest and a present whose objective is to reverse those practices of enclosure and dispossession. Calum MacLellan, spokesperson for Stòras Uibhist at its celebrations of the purchase of the South Uist Estate on St. Andrew's Day, 2006, known locally as "Independence Day", extended the geographical reach of these events. "This is where the New World is now!", he proclaimed. After 400 years of absentee landlordism, people of the

Highlands and Islands were no longer those "left behind" by the violences of the past and the migrations overseas that accompanied the Clearances of the eighteenth and nineteenth centuries, but at the forefront of initiatives whose intent was to work collectively towards more socially just and sustainable futures globally. Whereas the "New World" – as North America – was driven by an ethos of privatization, this new "New World" conjures a different politics, one marked by a narrative of a-commoning land.

In this and subsequent sectors of the chapter, I want to build a line of argument that suggests that this process of commoning the land through community land ownership disrupts a process of enclosure and privatization that culminated in the Clearances of the eighteenth and nineteenth centuries and reclaims the land through a collective, historically resonant, ethic that pre-dates that time. The community land movement, I argue, displaces a narrative of property characterized in broad terms by "the ownership model" (Singer, 2000) – complicated by feudal tenure though it may have been – and conjures in its place a counternarrative that removes the land as commodity from global property markets. As a distinctly counterhegemonic move, it strikes at the core of the process of neoliberalization (Mansfield, 2007a), reversing norms that have defined tenurial relations for hundreds of years and suggesting what Gibson-Graham (2003: 53) refers to as "a politics of the 'otherwise'", a politics of different possibilities. As I show, the process of de-commodification is congruent with Part 3 of the Land Reform (Scotland) Act 2003. In the process of reworking rights to the land, as the example of the North Harris Trust demonstrates, people's individual and collective subject positions *vis à vis* the land and each other change in complex and contingent ways.

I trace the process of commoning of the land – and the mobilization of property as always and actively "doing" (Blomley, 2003: 122) – through the metaphor of *dùthchas*, a term that may be conceived as both an inherited right and an evolving right to the land, to adapt Catherine Nash's (2002: 39) immensely insightful discussion of genealogical identities in Ireland. It is through this metaphor that past rights to the land may be narrated, that resistance to the deepening of the process of commodification of the land during the Clearances may be understood, and that the contemporary movement towards community land ownership may be charted. It calls into being collective rights and a collective identity rather than an individual right and identity; it suggests that these collective rights are open and changing rather than fixed, essentialist and exclusionary.

The exact historical beginnings of dùthchas are unclear but appear to be traceable to the social organization of the *tuath*, the mode of social organization in the Highlands and Islands that preceded the clan system (Hunter, 1995a). Characterized by Hunter (1995a: 58–59) as "a small, self-governing and largely self-contained community", the tuath, like the clan structure that followed, was far from being a democratic or egalitarian institution.[7] As with the clan, it was organized hierarchically through the rules of kinship so as to be readily mobilized in the case of hostilities. Its significance for the argument here is twofold. First, its law tracts, which, writes Hunter (1995a: 59), were "already ancient" when written down in Gaelic in the seventh and eighth centuries, included concepts of common property, specifically of woodland, non-cultivated land or hill pastures, and fish in the burns. The law codes specified, for example, the ways through which woodlands pertaining to the tuath were to be managed in the interests of long-term sustainability, fines being in place for any infringement (Hunter, 1995a: 63). "The notion that such resources ought to belong to the generality of people in the locality, rather than to privileged individuals", writes Hunter (1995a: 64), "is one that Highlanders have still not given up". This claim is captured in the still widely cited Gaelic proverb:

> *Breac a linne, slat a coille*
> *Is fiadh a fireach*
> *Mèirle anns nach do ghabh*
> *Gaidheal riamh nair* (Grant, 1961: 7)

(in translation: A Highlander has a right to take a salmon from the pool, a branch from the wood, and a deer from the hill).

Second, underpinning this claim to resources, was the notion of dùthchas, a word not readily translatable into English. Its roots, like those of *dùthaich* (land, territory), derive from an Indo-European word for "earth" (Newton, 2009: 306). For John MacInnes (2006: 279), dùthchas is "ancestral land or family land; it is also family tradition; and, equally, it is the hereditary qualities of an individual". Charles Withers (1988: 389, emphasis his) refers to the concept as "the expressed collective belief in the inalienability of the land; not in the sense of its formal appropriation through law as property or as a materially measurable commodity, but in the sense of land as *their* land, an inherited occupance, a physical setting with which Highlanders were indissolubly tied through continuity of social and material practices". Its meanings are captured by the words of a crofter on Harris, a man who traces his lineage to the times of Norse rule:

> There is no sense of ownership; it is a sense of belonging. You are part of the land. ... It is your heritage. ... In Gaelic, you never think about the land belonging to you; it is you that belongs to the land. The people belong to the land. That's the only connection that's made in relation to people and land. ... People belong to that land. ... Not just the land, but the whole concept of belonging to that land, everything that goes with the life we live here. ... These are inherited rights that nobody can argue with (interview, May 1997).

Its "flexible and fluid qualit[ies]", writes Michael Newton (2009: 307), were "useful" when "customs, values, beliefs and duties" were under threat.

Collective rights to the land, negotiated through the social practices of the tuath and then the clan, were reconstituted through everyday material practice. For centuries prior to the Clearances, the inbye land was worked collectively through the runrig system of cultivation. Comprising elongated mounds of soil, rigs, or *feannagan* in Gaelic, were built up from the soil of the intervening hollows (Brien, 1989: 151).[8] Manure and, along the coast, seaweed, were used to maintain fertility (Withers, 1988: 220–221). In areas of extremely limited agricultural potential, the feannagan might measure little more than the size of a dining table; elsewhere, they straddle a hill side. They were worked intensively, for the most part with the *cas-chrom*, the crooked spade, an implement specifically adapted to the local, rocky, conditions (see Dwelly, 1994: 172). The rigs were worked collectively but rights to the produce from the various strips were held individually, each household being allocated rigs of differing quality. The rigs were rotated on an annual basis for reasons both of equity and ensuring periods of fallow (Hunter, 1976: 113–114). The system thus fostered a collective ethic, made clear in the poignant testimony of a crofter from Geocrab, Bays of Harris, appearing before the Napier Commisssion in 1883. When asked to compare the current system where the inbye land was divided into individual lots and the practice of runrig had fallen into desuetude, Donald Morrison replied, "I have seen a woman weeping at being separated from her neighbours by the division of the crofts" (UK Government, 1884: 852). Inscribed in the land through collective human labour, the rigs appear today as bright stripes against the dark shadows of the intervening hollows in the sunlight of a winter's afternoon. They, together with the shielings, the rough grazings, where the women and children would tend the cattle during the summer, signalled the binding of people to the land.

Naming places – in Gaelic – was part and parcel of this binding, claiming the land through daily material practice. "The land speaks to us through our language", writes Ruairidh MacIlleathain (2007: 4). And frequently the names used are the same as those used for parts of the body – *sròn* translates as both "nose" and "ridge running off a mountain"; *cioch* refers to both "breast" and "breast shaped hill" (MacIlleathain, 2007: 12). "Every hillside, every lochan, every erratic rock", writes Roger Hutchinson (2003: 14), "told a story and had a name". The land was, and is for people with genealogical depth in a place, densely saturated with names, some stretching back through the generations, some reflecting the long Norse influence in these islands.[9] To Charles Jedrej and Mark Nuttall (1996: 123), the names are "mnemonic devices" that do not establish "a claim to possession" but, by being passed on orally through people's everyday interaction with each other and the land, produce "a sense of social continuity".

Whether prescribed through the tuath or the clan, claims to the land through dùthchas rested on the expectation that, in return for "allegiance, military service, tribute and rental" on the part of the "masses", the "ruling families had the responsibility to act as their protectors and guarantee secure possession of land" (Devine, 1994: 11). However, beginning in the twelfth century, at which time feudal law was introduced in Scotland (Pillai, 2012: 5), the expectations inherent in dùthchas – of reciprocal rights within a kin-based collectivity – were challenged by a legal regime based on quite different principles (Devine, 1994: 10).

As a system of land holding, feudal tenure was premised on the idea that land could have an "owner" and, as such, was alienable, both principles in direct contradiction to those underpinning dùthchas. Granted that ownership rights were not outright or allodial under feudalism (Callander, 1998: 9), it is nevertheless the case that this introduction of the notion of private ownership under feudal law laid the ground for the commodification of the land as the Highlands and Islands became more deeply caught up in the manoeuvrings of capital and discourses of "progress" and "improvement" in the eighteenth and nineteenth centuries (Hunter, 1976; Devine, 1994).

Under feudal tenure in Scotland, "the ultimate right in the land is vested in the Crown" and other claims to the land "are derived from the Crown either directly or through intermediate holders" (Gordon, 1999: 25). As William Gordon (1999: 25) explains, land is "granted out in consideration of services, real or nominal, to be rendered by the grantee to the granter". Rights are thus conditional rather than absolute, the granter maintaining "a legal interest in the land" (Gordon, 1999: 25). From the Crown, as Paramount Superior, at the apex, a pyramidal hierarchy of interests passes to superiors (granters) and from them to vassals (grantees) (Callander, 1998: 10). A superior holds interests that may, for example, include that of pre-emption, that is "the right of first offer to buy back a property if a vassal should decide to sell" (Callander, 1998: 10). A vassal holds conditional rights referred to as *dominium utile*, which Gordon (1999: 25) translates as "ownership (*dominium*) in an extended sense (*utile*)". These rights, as Callander (1998: 10) notes, "are subject to all the reservations and burdens [feus] of their feudal superiors". In turn, tenants derive their rights from the vassal (Callander, 1998: 10). The system becomes highly complex and rights become highly contingent, as there is no limit to the number of burdens that can be placed on a particular piece of land. "At each stage", writes Callander (1998: 10), "superiors can limit the extent of possession conveyed by reserving rights to themselves (for example, mineral rights) and by imposing additional burdens and conditions on the vassals (for example, that the new owner has to obtain the superior's permission to erect any new building or carry on any trade or business on the land)".

The right of individual ownership of the land under feudal law was further reinforced by legislation such as the 1695 Act Anent Land Lying Run-Rig which "allowed for the division of commonties[10] without majority agreement" (Pillai, 2012: 8). The Clan Act (The Highland Services Act 1715) and legislation enacted subsequent to Culloden – the Tenures Abolition Act 1746 and the Heritable Jurisdictions Act 1746 – supported the process of individual proprietorship (Combe, 2006: 197). The Heritable Jurisdictions Act, for example, recognized a clan chief as the sole proprietor, effectively rendering legally invisible rights accorded through dùthchas. In turn, the legislation supported the growing movement of enclosure and privatization, which led, in turn, to the forced removal of vast numbers of people in the interest of large-scale sheep rearing and, later, the establishment of sporting estates (Hunter, 1976; Devine, 1994; Wightman, 2010). The land's increasing value as a commodity clearly struck at the heart of a land holding system that recognized the inalienability of collective rights of the user to secure tenure as understood through dùthchas. Thus, while feudal tenure in Scotland was quite distinct from that pertaining in England and elsewhere,[11] its effect in terms of privileging private ownership rights – to exclude and to alienate – during the period of the Clearances paralleled those of the ownership model as theorized through Anglo-American law.

Subsequent to the enactment of legislation from 1885 onwards, and "with the notable exception of feudal burdens", argues Aylwin Pillai (2012: 6–7), feudalism became a spent force in Scotland. While this may indeed have been the case in general – that feudalism was more "a theoretical construct" than "a living system" (Law Society of Scotland, 1987–93, XVIII: 165, cited by Hunter, 1995b: 2) – it remained, argues Hunter (1995b: 2), "at the heart of Scotland's law". It was finally ended on 28 November 2004 under the Abolition of Feudal Tenure (Scotland) Act 2000, one of the first major pieces of legislation of the reinstated Scottish Parliament (Steven, 2004: 2).[12] Under this legislation, superiorities were ended and rights held by vassals became allodial or "absolute" (Steven 2004: 5).[13] This does not mean, or course, that such rights are uncomplicated by the "public" rights invested in the Crown or by neighbours' rights and tenants' rights (Callander, 1998: 27).

As the pace of enclosure and privatization quickened, the notion of dùthchas provided "the cultural force" (Devine, 1994: 11) that inspired resistance, culminating in the Highland Land Wars of the 1880s and 1890s (Hunter, 1976: Chapters 8–10). This is evident in the statements of witnesses appearing before the Napier Commission in 1883 and in the archival records of the Highland Land Law Reform Association, formed in 1883 and renamed the Highland Land League in September 1886. The latter's objective was the restoration "to the Highland people [of] their inherent rights in their native soil" (cited by Hunter, 1974: 57). Characterized by Hunter (1976: 154) as a social movement rather than a political party, its members – although not its leadership – were overwhelmingly comprised of crofters and cottars. It was at the forefront of mobilizing political protest both before and subsequent to the Crofters' Holdings (Scotland) Act 1886.[14]

Further evidence of the material and discursive force of dùthchas comes from the poets associated with crofter resistance to ongoing dispossession in the late nineteenth century. Central to the poetry of the 1870s, writes Meek (1976: 323), "lay the view that land was the property of the people, and not of any feudal superior" and, while "the people's right to the land was not written, it was considered to be beyond dispute". Foremost among these poets of the Land Agitation, as it was known, were Mary MacPherson or, as she was more popularly known, Màiri Mhór nan Oran (literally, Big Mary of the Songs) from Skye and Iain Mac a'Ghobhainn/John Smith, of Iarsiadar, Lewis. Referred to by

Hunter (1995a: 143) as "the great bard of the nineteenth-century Highland Land League", Màiri Mhór was passionately committed to crofter resistance. Through such poems as the well-known "*Eilean a'Cheo*" ("The Isle of the Mist" [i.e. Skye]) and "*Brosnachadh nan Gàidheal*" ("Incitement of the Gaels"), MacPherson exhorts her people to continue to fight for the land that is theirs (Meek, 1976: 322; Thomson, 1990: 245). "Her best work", writes John MacInnes (2006: 390), "gives the sharp feel of immediate experience while at the same time conveying the pressure of contemporary history". However, her work is analytically inconsistent. Samuel Maclean (1941: 319) draws attention, for example, to her reluctance in some poems – shared with a majority of other poets of the period – to recognize that it was not in fact "the English" who were forcing her people off the land. "[I]t was very plain", he writes (Maclean, 1941: 319), "that not one Clearance had been made in Skye by anyone who had not a name as Gaelic as her own". It was difficult for Highlanders, argues Meek (1976: 323), "to accept that those who were once their protectors were now their enemies and exploiters". On the other hand, as Maclean (1941: 323) remarks, MacPherson could acknowledge her "mistakes", and in "*Duilleag gu Gàidheil Chanada*" ("Message to the Gael in Canada"), she attributes blame to "the rack-renting, absentee landlords of Skye, who enjoyed in London the fruits of their exploitation of her people". In this poem, her work is closer to that of Iain Mac a'Ghobhainn, the most analytic of the poets of the time. His "*Oran Luchd an Spors*" ("Song for Sportsmen"), for instance, links the Clearances and the creation of sporting estates in the Highlands with the trade in opium in China (Maclean, 1941: 311; Meek, 1976: 315; Thomson, 1990: 240). Here, Derick Thomson (1990: 240) remarks, the poet is referring to James Matheson, then owner of the Isle of Lewis, whose firm, Jardine and Matheson, had business dealings in the east and whose profits from trading in opium were used to purchase that island in 1844. In this poem and elsewhere, Mac a'Ghobhainn recognizes capital's global circuitry, thereby setting himself apart from the great majority of the other poets of Highland resistance.

The Land Reform (Scotland) Act 2003 – Part 3

Reflecting in the Fifth John McEwen Memorial Lecture on Land Tenure in Scotland in 1998, a year before devolution and the re-establishment of the Scottish Parliament, then Secretary of State for Scotland Donald Dewar spoke of the opportunities for addressing land reform under a reinstated Parliament, something that had not been achievable in Westminster. "[T]here is undoubtedly a powerful symbolism", he said, in placing land reform at the forefront of legislative action (Dewar, 1998: 3). Emphasizing the need to ensure the public interest in land in any land reform, he maintained:

> It is clear that we need an integrated programme of land reform legislation – sweeping away outdated land laws, properly securing the public interest in land use and land ownership, increasing local involvement and accountability – to fit Scotland for the 21st century (Dewar, 1998: 6).

The legislation should "look forward" to the new century "rather than backwards to punish the sins of the 18th and 19th centuries" (Dewar, 1997: 7).

Dewar's address took place near the end of a decade which had seen the unprecedented actions of the Assynt crofters (1993) and the islanders on Eigg (1997) bring into community ownership the land on which they lived. Both sets of actions propelled the "unfinished business" of the "land question" (Cameron, 2001) into the political limelight and underscored once again the centrality of land as a signifier of national identity (McCrone, 1997). Nowhere is this clearer than in the public response to the Assynt crofters' campaign to raise money to purchase the 21 132 acre North Lochinver Estate, placed on the market when the absentee (Scandinavian) owners went into liquidation in 1992. The sale threatened to split the estate into seven parts (MacAskill, 1999: 40). In the case of Eigg, it was the notoriety of a succession of deeply unpopular landlords – most recently Keith Schellenberg, an English business person and, after 1995, Marlin Eckhart (alias "Maruma"), a German artist, who bought the island from him – that galvanized public support (Dressler, 1998).

It had also been a decade of significant political initiatives at the level of government supporting both land reform and community land ownership. Following a pledge made by the Scottish Labour Party with respect to land reform in their Manifesto for the General Election of 1 May 1997, the Land Reform Policy Group (LRPG), chaired by John Sewel, had been established in October 1997 (Dewar, 1998: 3). Its remit, as set out in the first of three reports, was

> to identify and assess proposals for land reform in rural Scotland, taking account of their cost, legislative and administrative implications and their likely impact on the social and economic development of rural communities and on the natural heritage (LRPG, 1998a: 1).

Land reform was needed, the group stated, "on the grounds of fairness, and to secure the public good" (LRPG, 1998a: 3). Rather than making the case for reform on historical grounds, the group drew on "the language of providing opportunities for individuals and communities" on the one hand and, on the other, environmental sustainability (Cameron, 2001: 101).[15] The idea of sustainability or, in the words of the first LRPG report, "sustainable development", was central (see LRPG, 1998a: 3). It is also the case, Ewen Cameron (2001: 98) suggests, that the purchase of the Orbost Estate in north-west Skye by Skye and Lochalsh Enterprise in 1997 with the objective of creating new crofts – however problematic this initiative proved to be – fed into this growing momentum for land reform.

Other measures, specifically targeted at community land ownership, whose beginnings occurred at this time, included the establishment of what was to become the highly influential Community Land Unit in Highlands and Islands Enterprise (HIE) in 1997 and a request to HIE to set up a Land Purchase Fund to finance community buyouts (Dewar, 1998: 4). The aim of the Community Land Unit was to provide support to communities in the sustainable management of community owned estates, and in that regard it has received very high acclaim (SQW, 2005). In 2001, the Community Land Unit and Highlands and Islands Enterprise, partnered with Scottish Enterprise, bid successfully to deliver the Scottish Land Fund/Ionmhas Fearann Na H-Alba. Initially established with £10.78 million from the New Opportunities Fund Lottery (later the Big Lottery), the Land Fund played a key role in supporting community land purchases in rural Scotland. (See Wightman, 2010: 149).[16]

In this context of convergence between a growing grassroots movement of community land ownership and Scottish Office initiative in the years leading up to the reinstated Parliament, and alongside other legislation pertaining to property (The Abolition of Feudal Tenure [Scotland] Act 2000, the Title Conditions [Scotland] Act 2003 and the Tenements [Scotland] Act 2004), land reform legislation was introduced. Subsequent to an extended period of consultation on, first, a White Paper (Scottish Executive, 1999b) and then a draft bill, the Land Reform (Scotland) Act 2003 became law, receiving royal assent on 25 February 2003. It comprises three parts. Part 1 concerns access rights; Parts 2 and 3 legislate for community land ownership under specific sets of circumstances. Part 2, The Community Right To Buy, which pertains to all of rural Scotland, establishes the terms under which communities can exercise a pre-emptive right to buy land. This means that communities, where they have registered an interest in the land, have the right to buy the land if or when it is placed on the market.[17] The more radical Part 3, The Crofting Community Right To Buy, details the terms according to which crofting communities have an absolute right to buy. This right is "an expropriative measure", meaning that a land owner's consent is not required for the sale of "eligible" land to an "eligible" crofting community (Macleod *et al.*, 2010: 97–98).[18]

My focus in this part of the chapter is on those sections of Part 3 of the Land Reform Act that allow me to pursue the following argument. First, I am concerned to trace the ways through which The Crofting Community Right To Buy disrupts the neoliberal norms of the ownership model of property. I argue that this community right to buy – by reinstating a common claim to the land, as captured by the metaphor of dùthchas – has the potential to remove land as a commodity from circulation in global capital markets and thus to reverse processes of enclosure and privatization. Second, discussion centres on matters of governance that are contained in Part 3 of the Act. I argue that by re-signifying the land in terms of collective rights and responsibilities, and by defining this collectivity or community in terms of place rather than interest, the Act conjures a politics of the possible that works against exclusionary claims and essentializing identities. It thus allows a more generous configuration of sustainability and social justice than is likely through private tenure. That this process of reworking the land/community may be contentious is evident.

Viewed in the light of current land reform in South Africa, Scotland's legislation has been characterized as a "modest response to a modest problem" (Reid and van der Merwe, 2004: 670). A comparison of the legislation in the two jurisdictions leads to "difference and not similarity", write legal experts Kenneth Reid and C.G. van der Merwe (2004: 670), despite a commitment in both places to the privileging of public over private interests in land legislation (Carey Miller and Combe, 2006: 24). Land reform in Scotland, argue David Carey Miller and Malcolm Combe (2006: 24), "remains unititular and focuses reform efforts at transferring ownership to a deserving community body within the traditional framework of Scots property law". In contrast, South Africa's land legislation, part and parcel of broader constitutional change, "appears to challenge … traditional notions of Roman–Dutch property law" (Carey Miller and Combe, 2006: 24). Notwithstanding these very generalized comments, Carey Miller and Combe (2006: 24) recognize in Part 3 of Scotland's land reform legislation "a fairly radical step away from the traditional protection afforded to Scotland's landowners". John MacAskill (2004: 129) goes further. Despite its shortcomings, he writes,

The crofting community right to buy in Part 3 of the Act certainly provides a reassuring example that the Scottish Parliament is taking its Viagra, and that it has not flinched in the face of the ECHR [European Convention on Human Rights], proprietorial and, indeed, crofting vested interests, and the Scottish Parliament has produced an historic piece of legislation (MacAskill, 2004: 129).[19]

What is not acknowledged here, or in other critiques (for example, Combe, 2006), is that the radicalness of Part 3 lies not simply in its tackling of the issue of social (and, for many, historical) injustice, but that this assault on class privilege is at the same time a troubling of the norms of property through which class interest is brokered. Land is, or has the potential to be, no longer a commodity with exchange value in global property markets. It is this threat to the commodification of land, to its value as a commodity tradable in global real estate markets and, through this, to the exercise of class-sanctioned "rights", that lay behind the sustained opposition to the process of land reform of the Scottish Landowners' Federation, captured so vividly by such phrases as "a Mugabe-style land-grab" on the front page of *The Daily Mail*, 24 January 2003.

To turn first to the land which is subject to The Crofting Community Right To Buy, the first two sections of Part 3 of the Land Reform Act, Sections 68 and 69, specify that a crofting community body may purchase all land under crofting tenure (inbye land, common grazings and land held runrig), salmon fishings and mineral rights without the owner's consent. The right to buy does not include the right to buy owner-occupied croft land (Section 68[3]), that is, land that has already been purchased by an individual under the Crofting Reform (Scotland) Act 1976.[20] Section 70 stipulates further that "eligible additional land" and "eligible sporting interests" may also be bought, subject to certain limitations, again without the land owner's consent (Section 77; see MacAskill, 2004: 111). The land in question must be "contiguous to the eligible croft land" that is the subject of the community body's purchase and owned by the same owner of that land (Section 70 [4] [a, b]). In both a situation where the owner does not wish to sell the land under crofting tenure and written application must be made to the Scottish Ministers (Section 73) and where the owner does not wish to sell "additional eligible land" and the case goes before the Land Court (Section 77), criteria for adjudicating the applications include the stipulation that the purchase be in the public interest (Section 74 [1] [n] and Section 77 [3] [a], respectively). Public interest is to be assessed on the grounds of essential community "development" and compatibility "with furthering the achievement of sustainable development" (Section 77 [3]; Section 74 [1] [j] and Section 77 [3] [b]). The concept of "sustainable development", which, MacAskill (2004: 121–122) observes, "lies at the heart of the Act", lacks a "workable definition", a failing he attributes to the government's concern that any definition would be open to legal challenge.

Section 71 of the Act identifies the legal body that may exercise the absolute right to buy. While the Justice 2 Committee, the lead committee examining the bill, considered a number of options in its deliberations of the draft bill and recommended that the government "consider a more flexible approach to community bodies" (Justice 2 Committee 2nd Report 2002 – the Stage 1 Report, Volume 1, para. 82-87, cited by MacAskill, 2004: 112), the Act specifies that the crofting community body be a company limited by guarantee (Section 71 [1]).[21]

Under this legal arrangement, the liability of its members is limited "by the amount which each member undertakes (guarantees) to pay" if the company is dissolved (MacAskill, 1999: 58). For MacAskill (1999: 58), this option provides a more robust legal basis for ensuring that the land remain in community control "in perpetuity" than for example, that of a company limited by shares. "With a company limited by share", he writes (MacAskill, 1999: 58), "each crofter would own shares which, without carefully drafted controls through a shareholders' agreement and the articles of association, could easily be sold to a third party". Drawing on the work of Andy Wightman (1996), he gives the example of the Glendale estate on Skye. As a company limited by shares, the members of Glendale Crofters' Ltd have the right to sell their shares on an individual basis. Wightman (1996: 181) writes that, "Many of the crofts have now been sold on the open market. The communal framework for decision making has been virtually destroyed" (also, see Hunter, 1991: 134). This situation he contrasts with that of the Assynt Crofters' Trust, a company limited by guarantee, where land owned by the trust may only "be disposed of on a collective basis". The one exception to this "rule", as Wightman (1996) notes, is the right of individual crofters, under the 1976 Act, to purchase their own crofts (see endnote 34), a provision at odds with the Land Reform Act 2003 and one that is a potential threat to community ownership (MacAskill, 2004: 131–132), a matter I consider later.

Part 3 of the Act details specific provisions that must be included in the memorandum and articles of a company limited by guarantee. These include the requirement that there be a minimum of 20 members, that the majority of members of the company be members of the crofting community, and that members of the company "who consist of members of the crofting community have control of the company" (Section 71 [1] [c, d, e]). There is thus, as MacAskill (2004: 112) observes, neither a requirement that membership of the company be restricted to members of the crofting community nor that tenants of crofts should form the majority interest. For the purposes of the Act, a crofting community is defined as follows. I quote in full from Section 71 (5) (a) as the matter is of particular importance in the development of the argument in the next part of the paper. A crofting community is defined:

(a)   as those persons who –
    (i) are resident in the crofting township which is situated in or otherwise associated with the croft land which the crofting community body has a right to buy under this part of the Act; or
    (ii) being tenants of crofts in that crofting township, are resident in any other place within sixteen kilometres of that township, and who are entitled to vote in local government elections in the polling district or districts in which that township or, as the case may be, that other place is situated; or
(b)   if, in Ministers' opinion, it is inappropriate so to define the crofting community, in such other way as Ministers approve for the purposes of this paragraph.

A crofting township is defined as:

(a)   any two or more crofts which share the right to use a common grazing together with that common grazing and any houses pertaining to or contiguous to those crofts or that common grazing; or
(b)   any combination of two or more crofting townships within that meaning (Section 71 [6]).

As a departure from earlier legislation – for example the Transfer of Crofting Estates (Scotland) 1997 Act, which applies to land held by the Scottish Ministers and which privileges crofters' rights over those of other residents (MacAskill, 2004: 114–115) – Part 3 of the Act clearly states that the crofting community is defined in terms of place rather than interest and thus is inclusive of both crofters and non-crofters. As MacAskill (2004) points out, it was this inclusiveness that had been favoured by Lord Sewel, as chair, and John Bryden, as external assessor, of the LRPG, and that had later been endorsed by the Justice 2 Committee (Stage 1 Report, Volume 1: 31, cited by MacAskill, 2004: 115). For Bryden (2007: 3), a place-based definition had the advantage of recognizing the potential social and economic contributions of non-crofters to the sustainability of local communities. It would also work against the ambitions of particular interest groups whose purchase of the land would run counter to the intent of the land reform legislation and the objectives of the local community.

The issue proved to be contentious. The Scottish Crofting Foundation (now Federation) argued for greater crofter control of the community body, including a "requirement" that crofters held a majority of the directorships (MacAskill, 2004: 116). John MacKenzie and Allan Macrae, both leaders in the Assynt Crofters' Trust, expressed dismay that the Act did not give the same rights to crofters as they had instituted in Assynt, namely collective crofter ownership of the land (MacAskill, 2004: 117).[22] However, the government's position was clear. As expressed by Ross Finnie, then Minister for the Environment and Rural Development,

> The croft tenants are usually a minority within crofting communities and have their own vested agricultural interests which can be at odds with the need of the wider community. There was therefore a risk that the effect would be to replace the landowner with what would simply be another monopoly interest. Furthermore, because the croft tenants already have considerable power to dictate the way the land is used the end result would be ownership by a vested interest with greater power than before (Justice Committee 2, Stage 1 Report, Volume 2: 373, cited by MacAskill, 2004: 117).

MacAskill (2004: 118) notes, however, that there are three places in the Act that, however imprecise the wording, provide openings where crofters could be accorded a measure of protection in the constituted crofting body. The first, located in Section 74 (1) (l), concerns the stipulation that, before consent is given to an application for a community land buyout, Ministers must be satisfied that the crofting community is "an appropriate crofting community". The explanatory notes indicate that this provision means that "the crofting body and the crofting community to which it relates [must] fully represent the crofting interests" in the land in question (Explanatory Notes to the Land Reform [Scotland] Act: 32, cited by MacAskill, 2004: 118). Second, Section 74 (1) (m) indicates that the right to buy may only be exercised if approved by the crofting community body. This approval is measured by a ballot, which, to succeed, requires both a simple majority of "those voting" and a simple majority of tenants of crofts who vote (Section 75 [1] [b] [i, ii]). As MacAskill (2004: 119) notes, no quorum is required; i.e., it is not necessary for a majority of crofters to be in favour of the buyout, just a majority of those who vote. It is also the case that, as the legislation uses the words "tenants of crofts", only the person legally designated as the tenant may count in this regard, not his or her spouse (MacAskill, 2004). As a third point, MacAskill (2004: 120) observes that in no place does the Act actually prohibit crofter control of the community body, whether as members or as directors, thus leaving open to democratic practice – under the bye-laws of the company limited by guarantee – the question of whose interests may prevail.

A commoning land

On only two occasions have communities applied to purchase land under the provisions of Part 3 of the Land Reform Act, The Crofting Community Right to Buy. In the case of Urras Oighreachd Ghabhsainn/Galson Estate Trust in North Lewis, the launch in 2005 of a "hostile bid" to purchase the 56 000 acre (22 662 ha) estate was sufficient to bring the owners back to the negotiating table and to reach a last minute "amicable agreement" with the community. In the case of the Pàirc Trust in South Lewis, the application to purchase that estate of 26 790 acre (10 846 ha) under the terms of Part 3 of the Act was finally given approval by the Scottish Ministers in March 2011.[23] Unfortunately, early celebration of the historic decision has subsequently been overshadowed by the land owner's decision to take the case to the courts (Chapter 4). In both cases, the owners had complicated the meanings of the land by invoking an interposed lease, a legal mechanism that would enable them, on sale of the land, to retain development rights. The mechanism worked through an owner's setting up of a shell company to which all development rights were hived off. Thereby, they exploited a loophole in the legislation, a matter that has since been redressed through the Crofting Reform (Scotland) Act 2007 (see Chapter 4). The matter was of substantial moment, as both communities' struggle to win the land was caught up in the struggle to control the wind, as a source of renewable energy (Mackenzie, 2006b). Given the degree to which claims to property in both cases have been interwoven with efforts to capture the wind, I leave further discussion to Chapter 4. Both cases demonstrate the degree to which the workings of property are played out in complex and frequently con-flictive ways as nature – here, the wind – is cajoled into commodification.

This limited formal engagement with the legislation, and its critics' assertion that substantial areas of the Highlands and Islands had been brought into community ownership without the legislation being necessary – The Assynt Crofters' Trust (1993), the Melness Crofters' Estate (1995), the Knoydart Foundation (1997), the Isle of Eigg Heritage Trust (1997) the Bhaltos Community Trust (1999) and the Isle of Gigha Heritage Trust (2002) are frequently cited in this regard – might lead to the view that the legislation has had limited impact. But a recent report (Macleod *et al.*, 2010: 132) confirms what many have long argued, that at least for those community buyouts that took place at the time of the public consultations of the Land Reform Policy Group and, later, the draft bill, as well as those that followed the Act's promulgation, the impending or enacted legislation provided a legitimating discourse for community purchase.

In the case of the North Harris Estate, where the community trust's purchase of the land took place in the period immediately before Parliament passed the Land Reform Act, negotiations were between a willing seller and a willing buyer. However, it has nevertheless been "conceded" by the selling agents, Knight Frank in Edinburgh, that the impending legislation had the effect of frightening off some prospective buyers (*WHFP*, 20 September 2002: 1). Buyers faced the uncertainty that they could, at any time, lose at least the land under crofting tenure through a community body's exercise of the absolute right to buy. On other occasions, the legislation has had the effect of encouraging private landlords to negotiate with community land trusts, Stòras Uibhist's threat of a hostile bid bringing the owners of the South Uist Estate to the negotiating table in 2006 being a case in point (Busby and Macleod, 2010: 601). Speaking on this particular buyout on the occasion of an international appeal for funds in London in May 2006, Brian Wilson insisted that "there is no doubt that [the mutual agreement] is underpinned by the legal right of crofting communities to buy their own land. That has crucially changed the balance of relationships" (cited in *WHFP*, 26 May 2006: 5). On these grounds, it may be argued that the Act and the deliberations that preceded it are of substantial material significance in altering the property map of the Highlands and Islands.

It is in light of the discursive significance of Part 3 of the Land Reform Act prior to 2003 that I now explore the ways through which a move towards the commoning of the land as it is brought into community ownership opens up the political possibilities of property to more sustainable and socially just configurations. Drawing primarily from the experiences of the North Harris Trust, I demonstrate how, through the "doing" or performance of community rights to land, people reposition themselves *vis à vis* the land and each other. I consider how the purchase of the North Harris Estate disturbed norms of property that had marked ownership practices for hundreds of years before focusing on matters of governance of the North Harris Trust.

For sale: The Amhuinnsuidhe and North Harris Estate

When Jonathan Bulmer placed what was then called the Amhuinnsuidhe and North Harris Estate on the market in 2002, community purchase of the estate was far from being a foregone conclusion. There was no assumption among the residents of North Harris that bringing land into community ownership was the unequivocal way to resolve problems of a declining population, outmigration of youth and limited job opportunities. The story of the buyout is about conversations through which the meanings of the land were reworked, conversations marked as much by caution and controversy as they were by excitement and a sense of justice.

Unlike such community buyouts as that of the Assynt Crofters' Trust and the Isle of Eigg Heritage Trust, where community action took place in a context of deeply unpopular landlords, Bulmer was generally considered to be "benevolent". He and his family lived on the estate and sent their children to the nearby school at Cliasmol; he employed local staff to run the estate; he hosted Christmas parties to which there was an open invitation. He was, nevertheless, but the most recent of a long line of landlords who, since the late eighteenth century, had exercised individual ownership rights over the land and exacted rents from the tenants of crofts, small though those rents at present may be.[24]

The land first entered the market as a commodity in 1779 when, congruent with a history repeated elsewhere in the Highlands and Islands, an indebted clan chief, in this case Roderick MacLeod of Dunvegan, sold the island to pay off debts incurred by conspicuous consumption in the south of England (Hunter, 2007: 5). Again, in common with experiences elsewhere in Scotland in the early years of the nineteenth century, Harris was caught up in the violences of the Clearances. "Improving" landlords or, on their instruction, their factors, forcibly removed people from the more fertile coastal land of west Harris and the strip of land between Huisinis and Kinlochresort and from

Cliasmol in north Harris in 1811, to make way for sheep farms (Hunter, 2007:5; also see Lawson, 2002: 140–173). The accounts of Harris crofters appearing before the Napier Commission in Obe (later, Leverburgh) in 1883 are instructive as to the degree to which acute poverty, hardship and social dislocation accompanied these forced relocations (UK Government, 1884: 845–864). While not all landlords may be tarred with the same brush as those involved in the Clearances – a few allocated land for resettlement in later years[25] – many residents of North Harris were, in 2002, aware that there was no guarantee that a future landlord would be as well-intentioned as Bulmer.

Bulmer placed the 55 000 acre Amhuinnsuidhe and North Harris Estate on the market on Thursday 25 April 2002 with a price tag of £4.5 million (MacSween, 2002: 1) – as a sporting estate. However sympathetic to the community's interest in purchasing the estate Bulmer later proved to be, the 28-page glossy sales brochure detailing the asset leaves no doubt as to its target audience (Knight Frank, 2002). The estate is produced, photographically, for the global, wealthy, propertied class. The front cover, in its entirety, is taken up by a photograph of Amhuinnsuidhe Castle, set in isolated splendour against the background of the "wild" Harris hills and, in the forefront, the blue of a calm sea. The estate, runs the text,

> must rank as one of the finest sporting estates in the islands of Scotland. There is salmon and sea trout fishing, the quality of which could scarcely be equalled on any other estate in the United Kingdom. The rugged countryside provides exciting stalking conditions with spectacular and beautiful scenery all around (Knight Frank, 2002).

The reader is then informed that the estate includes The Clisham (An Cliseam), the highest hill in the Outer Hebrides, and "a dramatic over hanging cliff" on Sron Ulladail. One of a number of climbing routes on the face, "The Scoop", is an E6 rock climb, described as "one of the finest routes in Britain" (Graeme Scott and Co., 2002: 51).

Text and photographs work together to create a "wild" landscape, empty of people. Except for one photograph of the sandy beach at Huisinis where several croft houses are visible, numerous photographs conjure the estate as uninhabited. This distinction is furthered by the listing of protective environmental designations that apply to the western part of the estate – as a Site of Special Scientific Interest, a UK designation, as a Special Protection Area and as a (then) candidate Special Area of Conservation, both EU classifications. The only people who appear in the brochure's pages are the ghillies exhibiting their catch of the day and stalkers dragging the carcase of a stag across the rocky ground or, through binoculars, scoping the landscape for their next trophy. As icons of wildness, salmon and deer are conscripted in a political technology (Foucault, 1979) that renders invisible others' claim to the land, giving to be seen only that which "fits" a land and landscape ordered through the claims of class privilege.

Silenced, except for a brief paragraph near the end of the brochure and the photograph identified above, are the individual and collective rights of 104 crofters to 20000 acres or 36 per cent of the land. This land under crofting tenure, as the modified 1:50000 Ordinance Survey map accompanying the sales brochure shows in diagonal hatching against a pink ground, runs as a ribbon around much of the outer perimeter of the estate.[26] Rents from these tenancies provide the estate with an annual income of approximately £355. But, the sales brochure's text assures the aspiring purchaser, "[t]he heartland of the estate, including the deer forest and main lochs, are [sic] not subject to crofting rights and are [sic] available with vacant possession". The map suggests an estate whose boundaries are fixed and resolute, where crofting interests are cartographically incarcerated, to adapt a phrase of Braun's (2002: 102), within a pink zone, and where a "heartland" remains, as from time immemorial, wild and untouched. It gestures towards "the sublime" (MacDonald, 2001).

In the brochure, the romance is furthered through the numerous photographs of the castle – opulence and splendour in every room – and a centrefold where, onto a background of sand, sea and sky and a horizon marked by island profiles, are projected the detailed plans of the castle's four floors, their layout premised on divisions of space consistent with the exercise of class interest. Built of imported sandstone in 1865 by Charles, seventh Earl of Dunmore, in the Scottish Baronial style, the reader is assured that the castle is "a modernised building of great comfort and charm without losing any of its original character". It is "the jewel in the crown" of the estate, to borrow a locally employed metaphor, central to the estate's definition, and the site from where the powers of ownership are extended (Photograph 2.2).

In view of the pivotal importance of the castle in the deliberations of community purchase over 100 years later, it is perhaps interesting to note that it was the extravagance of the seventh earl in building Amhuinnsuidhe that led to the break-up of the Harris estate, which at that time included the whole island (Hunter, 2007: 6). Finding the castle he had built at Aird a'Mhulaidh to be an unsatisfactory base for his

Photograph 2.2 Amhuinnsuidhe Castle.
Source: author.

hunting pursuits, the earl had built the much larger castle of
Amhuinnsuidhe (Hunter, 2007: 7). However, or so the story told locally
goes, his new wife, Gertrude, a daughter of the Earl of Leicester, was
unimpressed, likening the castle in size to her father's stables. Not to be
outdone, her husband added another wing to the castle, a move that
pushed him into bankruptcy and the sale of North Harris to Sir Edward
Scott, a London banker, in 1868 (see Hunter, 2007: 7–8).

Through photographs, the centrefold, maps and text, castle and
estate are produced in the brochure as one indivisible and indissolu-
ble property, the one a necessary part of the other. "Home" (the
castle) is tied to "heartland" through the configuration of nature as
"wild", the promise of trophies of deer and salmon the perquisite of
the landed class and their associates. The unity of property is thus
achieved through the conceptual separation of the "social" from the
"natural". The "wilding" of the land through protective environmental
designations contributes to this particular way of seeing, of normalizing particular practices of property. Conjuring property as
unchanging and unquestioned, the sales brochure becomes part and
parcel of an optic that deletes from the visible register any threat to

the norms of private land ownership. These threats had become increasingly visible in the intense debates about property that were reaching their zenith in the public consultations concerning the proposed land reform legislation at precisely the time the estate was placed on the market.

Troubling property – "The quiet revolution"[27]

The first publicly visible signs of disturbance of this model of land ownership in North Harris came at a meeting held on Monday, 29 April 2002, in the Tairbeart Community Centre four days after the sale of the North Harris Estate was announced. Donald MacLennan, local chair of the Scottish Crofting Foundation, had contacted the two Councillors for Harris on Comhairle nan Eilean Siar/Western Isles Council, Morag Munro and Donald MacDonald, and the local Member of the Scottish Parliament, Alasdair Morrison, regarding a meeting to which it was agreed should be invited all residents of the estate, not only crofters, and key people from those agencies who could provide information. John Watt, head of the Community Land Unit of Highlands and Islands Enterprise, and Simon Fraser, well-respected solicitor concerned with crofting community buyouts, were among these. About 40 people attended, a turnout considered to be good in view of the short notice given for the meeting.[28]

The economic viability of the proposal to buy the estate was the central concern expressed at the meeting. Among the divergent views given voice – in an exploratory rather than confrontational way – concern about taking on ownership and management of Amhuinnsuidhe Castle as part of the estate was clear (interview, member of the steering group, 27 September 2002). John Murdo Morrison, well-respected business man and opinion leader on the island, let it be known that his own experience in maintaining the Harris Hotel, also dating from 1865, led him to believe that the castle could prove to be "a millstone around the neck" of the community (cited by Hunter, 2007: 23). He opposed community purchase of the estate. Councillor Morag Munro was more sanguine. She spoke of the "huge challenge" the community faced: "Undoubtedly, people will have to look thoroughly at the viability of such an undertaking and weigh up the options. I would like to see it happening. Our hearts say go, but our heads say be careful" (cited by MacSween, 2002: 3). Focusing in this meeting on economic feasibility, she nevertheless referred in a later conversation to the debt she owed her ancestors who had fought for the land, recalling the words of a song written by Màiri Mhór nan Oran, poet of the 19th century Land League:

> *Mura toir sibh buille chruaidh*
> *Fhad's a tha 'n tuagh 'na ur làimbh*
> *Cumaidh iad sibh dol mun cuairt*
> *'Na ur truaghain gu bràth*

(In translation: "Strike a blow while the axe is in your hand or they [the landlords] will have you going round and round as poor souls for them for ever") (interview, 22 September 2002). Clearly for her, and others in the room, there lay a genealogically deep claim to the land on the grounds of an inherited and unalienable right understood in the term dùthchas. In this light, it is perhaps ironic to note that the one crofter at the meeting of 29 April 2002 who did draw on history to legitimate contemporary action – a man of relatively recent arrival on the island – was met with the response, from another crofter of much greater genealogical depth in North Harris, that, "We can't trade on what's happened in the past. We have to present this as a business plan" (interview, 30 April 2002).

Notwithstanding considerable scepticism evident at the meeting, the consensus was to appoint a steering group to explore the matter further by commissioning a feasibility study and to take responsibility for ensuring ongoing community consultation. Harris Development Limited, a community-led initiative on the island (Mackenzie, 2001), provided leadership experience and logistical support.[29] In order to promote wider community consultation, the steering group called a second meeting for 13 May 2002, attended by over double the number at the earlier meeting and, at that meeting, increased the number of people on the steering group to ensure broader community representation and "balance" – in the sense of being "non-political", of not being propelled by the agenda of a particular political party (interview, member of the steering group, 19 March 2004). Concern had already been expressed at this meeting by John Murdo Morrison that the people of North Harris were being used "as pawns in a bigger political game" (cited in *WHFP*, 17 May 2002: 1), reference being made to the land reform agenda of the local Member of the Scottish Parliament's party, Scottish Labour.

What was understood by the idea of "community" was confirmed at this second meeting. Membership was to be open to all adult residents of the estate, not just crofters. Harris crofters were well aware that the Assynt crofters had decided to restrict membership to those registered as tenants of crofts. Indeed, John MacKenzie, one of the leaders of the Assynt Crofters' Trust who had progressed that buyout, spoke at this meeting, impressing those present with his honesty. "He didn't try to paper over any cracks", recalled one member of the steering group (interview, 1 October 2002). However, with a leadership (and a solicitor, Simon Fraser) committed to an inclusive, place-based, definition of community and the knowledge that public funding for purchase of the land (through the Scottish Land Fund and the Community Land Unit of Highlands and Islands Enterprise) was conditional on such a definition, pragmatism prevailed. As one crofter expressed it, "[Simon Fraser] reckoned from the logistic point of view that all these people [residents] were part of the community. ... Whatever decision he comes up with, I'll go along with. [It] doesn't mean I agree with him" (interview, 3 October 2002).

During the period of community consultation that followed the meeting of 13 May 2002, it quickly became apparent that a counter-imaginary to land as commodity circulating in global property markets was tied up with tackling how the estate was to be defined. Debate centred on the castle of Amhuinnsuidhe: was it, to borrow locally enlisted metaphors, "the jewel in the crown" or "a thorn in the side?". Opinion was divided. There were certainly doubts about the lack of community experience in running the estate, with or without the castle, as the consultants, Graeme Scott and Co., learned from the series of meetings they arranged with different groups of residents, including students attending Sir E. Scott School. However, concern about the community's ability to turn a profit from the castle was particularly acute (Graeme Scott and Co., 2002: 23–27). One crofter with whom I spoke, who said she found it emotionally "hard to split the estate", was able to resolve her dilemma when she was asked the question, "What is it we want to control?". Her reply was, "Whatever we do, the community must have control. If the community has control of the land, [it] can control anything that happens to them. Control of the castle is neither here nor there as far as the community is concerned" (interview, 27 September 2002).

Another crofter who had worked in the castle as a maid during her summer holidays from college in the late 1960s and early 1970s emphasized the profundity of the significance of severing the castle from the land in terms of class but also of gender. Referring to a long line of landlords, including then owner Sir Hereward Wake, she stated, "One didn't question. That was the way it was. The land was the land – and we were there to earn a few pennies". "They didn't do anything for themselves" – a fire might have been set, but "they'd ring for someone [a maid] to come and light it for them", she said. She contrasted her poor wages with the wealth of the owners. It was "a disgrace to collect family support", she said, but that was necessary when you worked at the castle. The privileges of class and the workings of gender were even more marked in the 1930s when her mother worked at the castle. The landlord, Sir Samuel Scott, was regarded as "the almighty" and the butler as "almost the almighty". Lady Scott was similarly accorded a lofty position. When she came in the front door, "she'd whistle – a sign that everyone had to disappear, perhaps into the broom cupboard, so that she didn't have to see servants. There was a real hierarchy", she continued, "ladies' maids washed ladies' clothes and their own [i.e. the ladies' maids'] clothes were washed by 'serfs' [such as her] from the village". Most of those employed in the castle during its seasonal occupancy between the beginning of June and the end of August came from outwith the island. Only ghillies and a few "lower order" maids such as herself were locally employed. However, she continued, it was her view, that,

> the land should be for the people not a person who had mega-bucks. ... Crofters and land, it's very emotive. ... [Land] is the one constant. Everything else can vanish – the houses, ourselves – it will be here. ... Highlanders have a special feeling for the land – to do with history, being booted off it. ... You're close to the land, they've worked the land, ... needed the land in a way that city people don't have. And all this comes into the buyout. If the whole thing falls flat, I feel that we've achieved this. The land will remain, that's the bottom line (interview, 21 August 2003).

She strongly endorsed the separation of the castle from the land in the community's offer of purchase.

Matters came to a climax at the public meeting of 9 September 2002 when the consultants, Graeme Scott and Co., presented their report, *The North Harris Estate Feasibility Study*. As required by the terms of reference from the steering group, they had considered three main options: Option 1 considered a sale on the basis that the estate remain intact; Option 2 depended on severing the castle from the estate subsequent to the purchase of the entire estate; and Option 3 considered the purchase of only the land under crofting tenure by a "Crofting Community Group" (Graeme Scott and Co., 2002: 57). They recommended a variation of Option 2: selling the castle and other estate properties, together with salmon and sea trout rights, to a third party commercial operator, subsequent to the purchase of the whole estate (Graeme Scott and Co., 2002: 57). They considered it "ideal ... if the third party partner could be found in advance, so that a joint bid could be made" (Graeme Scott and

Co., 2002: 57). Aware of the degree of unease in the community with any proposition that included the purchase of the castle, the steering group recommended what it called a variation of this option, namely, the purchase of the estate, shooting and other (such as mineral) rights, rental properties, various buildings, a boat shed and slip, and excluding the castle, the salmon fishing rights, and the houses at Tolmachan and Kinlochresort that were connected with the fishing.

The profound shift in conceptualizing the estate – of severing the land from the castle – evident in the steering group's recommendation may indeed have been premised on a sound economic rationale. That is the case that David Cameron made on behalf of the steering group on 9 September (North Harris Steering Group, 2002). Moreover, that is the reason John Murdo Morrison reversed his earlier position. He had spent "many sleepless nights" worrying about the inclusion of the castle, he said, but now the proposition was "really worthy of looking at" (words recalled by interviewee, 27 September 2002). He was in favour. Spontaneous applause greeted his pronouncement and excitement replaced the scepticism that many present in the Hall still harboured. "A number of people came into the Hall against [the proposed buyout], who went out for it", observed one woman. She knew of a particular group of crofters who had been "dead against it – the resistance to change mentality. ... They had reasonable landlords down there ... and who's going to look after it". But, she continued, as the meeting progressed, first one and then the others changed their minds and "they were out canvassing on behalf [of the buyout]. It was amazing really in such a short time" (interview 27 September 2002).

A postal ballot counted on 17 September 2002 confirmed the sea change in public perception: 74 per cent of those eligible to vote (401 of an electorate of 539) cast a vote, of whom 75.3 per cent (302) were in favour and 24.7 per cent (99) against. With this mandate from the community, the North Harris Trust was created as a company limited by guarantee to take forward the community's aspirations. And, after the rejection of a first bid on the grounds that it was "not acceptable in its current form" – the seller wished to sell the undivided estate to one buyer or to two or more partners or, if it were to be divided, to ensure that further assets were attached to the castle – a second joint, but legally separate, bid with Ian Scarr Hall, a business man with long association of the island, succeeded.[30] The North Harris Trust now owned the North Harris Estate and a series of attached rights. These included sporting and brown-trout fishing rights, the Quayside cottages, the jetty and estate houses, rental assets including radio masts and wayleaves, the Amhuinnsuidhe fish hatchery site and most of the foreshore. Scarr Hall owned what was now called the Amhuinnsuidhe Castle Estate, which included the castle and surrounding 600 acres, salmon fishing rights west of the Stulaval to Uisgneabhal watershed (with the exception of Loch Gobhaig, which lies in the middle of a crofting township), the houses associated with them at Tolmachan and Kinlochresort, the stable block houses, the boathouse and the slipway. A partnership agreement was negotiated to cover rights of access and the management of specific assets. On 21 March 2003, the community took possession of the North Harris Estate.[31]

On 3 March 2006, and with considerable celebration, the 7500 acre (3036 ha) Loch Seaforth Estate, 99 per cent of which is under crofting tenure, was "reunited" with the North Harris Estate. It had been excised from the North Harris Estate when Hélène Panchaud, widow of Swiss businessman Gerald Panchaud, had sold that estate to Bulmer in 1994, owing to the perceived value of its mineral rights, from which she sought to profit.[32] The total estate now comprises about 25 304 ha, of which 11 862 ha (46.9 per cent) are under crofting tenure and 13 441 ha (53.1 per cent) are classed as a deer forest (SNH, 2007: 5; NHT, 2007: 18) (Map 2.1).

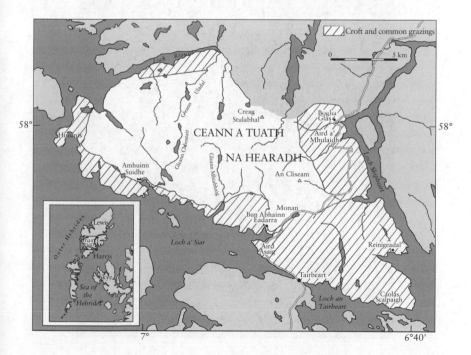

Map 2.1 The North Harris Estate/Ceann a Tuath na Hearadh. Map by Christine Earl, 2006, updated by Adèle Michon, 2011. Based in part on information from a map that the seller's agents (Knight Frank) included in the material advertising the North Harris Estate when it was placed on the market in 2002.

As Calum MacKay stated in the handing-over ceremony of the North Harris Estate cited at the beginning of the chapter, the occasions were historic. After centuries of private ownership, rights to the land were now in the hands of the community and no longer subject to the vagaries of the global property market. The norms of property – of enclosure and privatization – had been reversed and a process of commoning the land was underway (Photograph 2.3). Questioning

Photograph 2.3   Welcome sign to the North Harris Estate.
Source: author.

the spatial integrity of the estate – that the castle belonged to the estate and vice versa – denaturalized the legal and social premise of the ways in which the estate was given to be seen and provided other ways of seeing. As Blomley (2003: 134) has observed, "Property regimes can easily appear to be part of the landscape ... and as such their violences can appear to be the order of things". It was this ordering that was upset. The transfer of land to the community removed the most egregious signifier of class privilege, the castle, from the visual field, replacing an individual right of ownership with the complexities of collective rights of access, use and control. The detailed maps drawn up during the protracted negotiations provide evidence of this.[33] There is now, for example, a border between the trust's land and the Amhuinnsuidhe Castle Estate, signifying an end both to the norms of private ownership and to the relationship between a crofter as tenant and a private landlord. The boundary demarcating land under crofting tenure remains, such land being subject to separate crofting legislation, but its meanings have shifted. There is not simply a reversal of the situation where the tenant of a croft becomes landlord. Instead, as MacAskill (2004: 131) notes, a crofter now occupies a potentially contradictory position – as both tenant of a croft and, as a member of a community land owning trust, landlord. As the community is defined by residence rather than interest, congruent with the definition contained in the Land Reform Act, the crofter shares ownership not only with other crofters but also with those who are not crofters. That this can lead to conflict is evident in the case of Stòras Uibhist, which I discuss shortly.

The one caveat to this scheme of things – and a card played by the Assynt crofters in their campaign to buy the North Lochinver Estate (MacAskill, 1999: 97–128) – is the provision under the 1993 Act (first contained in the 1976 Act) for an individual right to buy. This provision allows a crofter to purchase his or her croft (inbye land together with an apportionment of the common grazings) at 15 times the annual fair rent – "a fraction of the price", as MacAskill (2004: 131) observes, that the community would pay for the land. His assessment is that, therefore, "the main asset of the crofting community body is, potentially, a dwindling asset" (MacAskill, 2004: 131).[34] This is obviously of more serious concern where all, or almost all, land is under crofting tenure. On the North Harris Estate, five of the 129 crofts are owner-occupied (interview, NHT Administrator, 1 April 2011).[35]

Governing places

Whether for the exercise of the right to buy under Part 3 of the Land Reform Act or for an application for funding from the Scottish Land Fund and the Community Land Unit of HIE, a community is required to register as a company limited by guarantee and to define its membership with respect to place rather than interest. These provisions are contained in its memorandum, articles of association, and bye-laws, which, together, define the community trust's objectives, how it is to be governed and who is eligible to be a member. Membership eligibility and structures of governance, I show here, provide the discursive spaces where the reworking of a collective subjectivity takes place. As detailed in the Memorandum and Articles of Association of the North Harris Trust (NHT, 2002), the objectives define a broad, socially and environmentally conscious, remit:

1. to take all appropriate measures to conserve the natural heritage ... of North Harris for the benefit of the community and public at large and to promote open public access thereto insofar as this is not detrimental to this conservation;
2. to promote trade and industry for the benefit of the general public;
3. to relieve poverty and provide help for the aged, handicapped and infirm and to advance education and other charitable purposes beneficial to the community;
4. to provide and promote the provision of housing for people in necessitous circumstances and also specially designed or adapted housing as may be required for the elderly, handicapped or disabled;
5. to develop or promote the development of infrastructure for the benefit of the general public to improve communications throughout North Harris.

The eligibility criteria for membership in the company as laid out in the bye-laws, amended in 2006 to reflect the inclusion of the former Loch Seaforth Estate, emphasize the necessity for residency (NHT, 2006). All individuals over the age of 18 (in the cases of the West Harris Crofting Trust[36] and the Galson Trust, the minimum age for membership is set at 16; for Stòras Uibhist it is 18) are eligible to be members provided that they reside on the estate, that they reside elsewhere in Harris or Lewis but are the registered tenant of a croft on the North Harris Estate, or that they reside elsewhere but are the registered tenant of a croft within the estate and "actively" work that croft. A resident is defined as "actually residing" on the estate for a minimum of nine months a year, except in the case where the individual is absent from their "permanent home" for reasons of education or because they are employed in the armed forces or the merchant navy. A lifetime membership fee costs £1.00. In April 2011, membership stood at 441.[37]

The significance of these criteria for membership in the land owning trust with respect to the negotiation of community lies in their opening up of rights to the land to residents who had previously not had these rights and in re-defining crofters' rights such that, as previously mentioned, crofters simultaneously occupy the position of tenants (of land under crofting tenure, as they did prior to community ownership) and owners of the entire estate, a position they share with non-crofters. Excluded from membership are those owners of holiday homes who reside on the estate for only a few weeks a year, a matter of particular moment in places such as West Harris, where 41 per cent of the housing stock (89 houses) on land now owned by the West Harris Crofting Trust is classified as holiday homes or self-catering cottages (CIB Services, 2008b: 2). With reference to the argument I am making, there is simultaneously a disruption of the norms of private ownership with community purchase of estates and a troubling of the boundedness or the "coherence" (Butler, 2004: 216) of the categories of owner and tenant as people reposition themselves *vis à vis* the land and each other. Through this process, property norms are disturbed and rights to the land are shown to be "malleable" and "transformable" (Butler, 2004: 216). These rights are claimed not through "sharp boundaries" and "fixity" but through an "edge politics" characterized by "overlap, layering, and movement" (Blomley, 2004: 155).

Community trusts' structure of governance, as stipulated in the Articles of Association, contributes additionally to the remaking of a collective subjectivity. As a company limited by guarantee, all members are eligible to be elected as a director, elections taking place at the annual general meeting. In the case of the North Harris Trust, where, with the addition of the Loch Seaforth Estate, the number of directors increased from nine to ten, three or four directors stand for re-election or are replaced, on a rotational basis, each year. Each director represents a geographical area, four being elected from Tairbeart, the largest settlement in North Harris, and six from crofting townships.[38] That crofters continue to predominate numerically as directors in North Harris does not constitute limited representation of difference. As Sharon Macdonald (1997) has shown on Skye, crofters differ in age, gender, "localness", "social standing" and "religious status", for example. In North Harris, the number of women elected to a directorship has varied between two and four.

These governance practices provide the discursive ground through which people rework their individual and collective subject positions. The re-articulation of people's rights to the land is central to this reworking. The North Harris Trust claimed land on the basis of a collective right, historically resonant with the notion of dùthchas, a principle that, as I have shown, had provided the legitimating force behind resistance to the enclosure and privatization of land that marked the Clearances of the nineteenth century. Consistent with the provisions of Part 3 of the draft Land Reform Bill, which was the subject of broad public debate at the time of purchase, this invocation of a collective right was, however, based on an inclusive notion of community defined through place – where one resided – rather than through a fixed marker of identity – for example, a crofter with or without genealogical depth on the island. Boundaries of belonging were thus disturbed. The meanings of the land, which remains the central signifier of identity in the Highlands and Islands (Hunter, 1976, 1991; Devine, 1994; McCrone, 1997; Hutchinson, 2003; Mackenzie, 2006a), shifted and dùthchas, an inalienable right to the land, was mobilized as both "inherited and evolving" (Nash, 2002: 39).

In the case of the North Harris Trust, due in large measure to the leadership's insistence that the membership be fully consulted and, through polls, be part of the decision-making process for any project of significance, there has been a lack of conflict between the leadership and members. For Stòras Uibhist, however, the story has been quite different. Here, the reconstitution of community was marked, at least until the spring of 2009, by ongoing and, frequently, bitter dispute. The leadership's proposal to refurbish the golf course at Askernish provides one widely publicized instance of this.

In November 2006, Stòras Uibhist, as the community land owning body is named, brought the 93 000 acre South Uist Estates, comprising almost all of Benbecula, South Uist and Eriskay, into community ownership. Among a raft of projects aimed at reversing population decline and economic stagnation, the restoration of the golf course at Askernish to its previous extent ranked highly. Run by the Askernish

Golf Club, the course had been singled out by John Garrity of *Sports Illustrated* as "the finest in the world", ranked ahead of Ballybunion, Ireland, in second place, and Pebble Beach, California, USA, in third (*WHFP*, 6 July 2007: 3). It was designed by nineteenth century Scottish "golf pioneer" "Old" Tom Morris in 1891 but, unlike other courses he planned, Askernish had not been altered to accommodate "the modern game" (*WHFP*, 29 August 2008: 22–23; *Stornoway Gazette*, 21 May 2009: 7). This Stòras Uibhist set out to do, backing the plans of the golf club to restore the course to 18 holes from the 9 that had been in play since the 1930s (*WHFP*, 6 March 2009: 2). Claiming that these plans had been agreed to by the previous owners of the estate (*WHFP*, 13 June 2008: 2), Stòras's leadership proceeded with the expansion, thereby setting itself against the crofters on whose common grazings the changes were to be made.[39]

Basing their case on the grounds that their rights as crofters were being violated and that "the spirit" of a document written in 1922 by the Board of Agriculture for Scotland and endorsed by the Scottish Land Court was being contravened, the Askernish Common Grazings Committee took their case to the Land Court in April 2007 (Bell, 2007: 1). In doing so, they acted on the basis of the majority, but not all, of their members. Of 11 shareholders, 7 opposed the extension.[40] The resignations of two Stòras board members in May 2007 on the grounds of disagreements over governance, including how the refurbishment of the golf course was being handled, provide an indication of how divisive the issue was in the community as a whole. The split was not just among crofters, who were now contradictorily positioned as both tenants and, together with non-crofters, collective owners of the estate, but also among the leadership.[41]

In February 2009, the Land Court ruled in Stòras Uibhist's favour insofar as it upheld the legal basis for the golf course and Stòras's entitlement to exercise that right (*WHFP*, 6 March 2009: 2). There was no restriction, the Court found, of the area within the common grazings where golf could be played (*WHFP*, 6 March 2009: 2). The golf course, deputy chair of the Court, Sheriff Roddy Macleod, confirmed, "had always been intended to co-exist with the crofters' rights to graze their soumings on the common grazings" (*WHFP*, 6 March 2009: 2).[42] However, he continued, "while the right to play golf was 'without geographical limitation', it 'must not have the effect of rendering that [the grazing right] impractical'" (cited in *WHFP*, 6 March 2009: 2). A second hearing would be necessary, Macleod maintained, to resolve that issue, barring an earlier out-of-court settlement. Such a settlement was finally achieved in May 2009, shortly before the date set for the second hearing.

Conclusion

Post Legislative Scrutiny of the Land Reform (Scotland) Act 2003 by Calum Macleod *et al.* (2010), insofar as it concerns Part 3 of the Act, highlighted a number of concerns, including the complexity of the application process, that it was "resource intensive and exhausting" (Macleod *et al.*, 2010: 132), part of which had to do with the onerous requirements for mapping. It was this requirement which led to the Galson Trust's and the Pàirc Trust's decisions to apply, in the first instance, to purchase only the land designated as common grazings. Other "barriers" to applications to buy the land under Part 3 of the Act were identified as a lack of funding and advisory support – in contrast to the early days of crofting community buyouts – and lack of "fit" with crofting legislation. With respect to the lack of fit, I have already identified the contradiction between, on the one hand, crofting legislation (initially, the Act of 1976 and, subsequently, the Act of 1993), which promotes individual ownership, and the Land Reform Act (2003), which legislates for community ownership. The two pieces of legislation work against each other, the one subject to market forces, the other working against the marketization of the land. As MacAskill (2004: 131) observes, this contradiction threatens the material base of community ownership.

That said – and there is clearly a case for amendments to the Land Reform Act if it is to fulfil its radical potential as the (now) majority Scottish National Party Government recognized in its manifesto for the May 2011 election – there is some evidence that Part 3 of the Act is beginning to have an impact on the property market. A "two-tier land market" appears to be emerging: those estates without land under crofting tenure are valued substantially higher than those with croft land as, in the former, there is no threat from eligible crofting communities wishing to exercise an expropriative right (Macleod *et al.*, 2010: 116). It is the land either wholly or in part under crofting tenure that has the potential to be removed from the global property market under the existing legislation, and it is these crofting estates that hold the possibility of troubling the norms of private property and, through a commoning of the land, making visible an alternative optic to that of commodification. It is here that a counterhegemonic commons of which McCarthy (2005a) writes may be created.

It is also the case, as I have demonstrated and as Macleod *et al.* (2010) endorse in their study, that as yet the major impact of the legislation has been through the creation of a legitimating discourse for a crofting community's right to buy rather than through the exercise of the right to expropriate. As Murdo Morrison and Brian Wilson made clear at the ceremony of giving sasine on the North Harris Estate on 21 March 2003, this authorizing discourse centres on the reclamation of an ethical and political commons (Blomley, 2008: 318) sanctioned through the historically resilient notion of dùthchas. Dùthchas, to follow Blomley's (2008: 318) theorizing of the commons, provides "a language of rights and justice" through which an ethical and political claim to the land is "justified and enacted". As metaphor, dùthchas interrupts property norms instigated under feudal tenure and deepened during the eighteenth and subsequent centuries as land became a commodity traded in global property markets. It counters an ongoing process of enclosure and privatization by making visible an historically resonant counternarrative to the ownership model of property. This narrative replaces "the power to exclude" – "to displace, evict and remove" – which is at the centre of the ownership model (Blomley, 2008: 316) with a power to negotiate inclusion. Injustice – of the past or present – is addressed not by subdividing the land and issuing rights of private (exclusive) ownership to those who can indelibly document their historical occupancy of the land, but by mobilizing a discursive space where material, political, and ethical rights to land held in common are negotiated as people rework their individual and collective subject positions. The practices of governance of community trusts and the bye-laws through which membership is defined in terms of place rather than interest are key in this regard. Discursively, they open a space where categories of belonging are destabilized and become more mobile and mutable. In Butler's (2004: 216) words, "the coherence of categories [is] put into question".

I now go on to show how this reworking of property and the co-constitution of a collective subjectivity – an alternative politics of property – allows the opening of the land to different material and metaphorical inscriptions of nature. Chapter 3 focuses on the "wild" and Chapter 4 on the wind. Each instance of the reworking of property troubles an optic that would conjure nature in the interest of private or corporate capital or the purview of outside conservationists. Each replaces an optic of an empty land with one that makes visible a nature that is constantly and continually a-doing.

Notes

1   The handing over ceremony was preceded, briefly, by the "opening" of the Millennium Forest on common grazings at Scaladail. The project, which involved the planting of "native" trees, was spearheaded by Murdo Morrison. (See Chapter 3.)

2   Unless otherwise indicated, the information provided here comes from *A Day of Celebration*, the North Harris Trust's video recording of the day's events.

3   I have suggested elsewhere (Mackenzie, 2006a: 594) that Simon Fraser reworked the meaning of the legal practice of giving sasine in the ceremonies on Gigha and on North Harris. In contrast to earlier practice under feudal tenure, the recent events were staged as a collective rather than individual right/rite (Mackenzie, 2006a: 594; see also Gordon, 1999: 26, 36–38).

4   In translation from the Gaelic: "All of us gathered together at this moment pray that you and every generation to come will prosper, and that you will receive every blessing on this land which our Lord has given you to enjoy" (Hunter, 2007: 99).

5   A "deer forest" refers not to a forest as such; trees disappeared from this area long ago (Smout *et al.*, 2005). The term had been used to denote the land on which the deer graze. The raid drew attention to one of the deficiencies of the 1886 Act which legislated rights for crofters but ignored the needs of those who were without land, the cottars.

6   Joni Buchanan (1996: 39–69) provides the most complete account of the Pàirc deer raid. The men from Lochs who raided that estate were for the most part impoverished cottars. Despite repeated petitions to the owner, Lady Matheson, they had been refused land on what had become a sporting estate used by the wealthy for the occasional shooting of deer. The land had been cleared of people in the first half of the nineteenth century. Buchanan (1996: 48–49) notes that about 200 men raided the estate on 22 November 1887, led by pipers, and having notified in advance the "authorities" and press of their intentions. She comments "The people of Lochs had, in a superbly effective manner, put their case to the nation – a fact reflected in the massive media coverage and subsequent Parliamentary attention given to the raid. They had been pushed to the abyss of poverty and injustice and had fought back in the only way open to them" (Buchanan, 1996: 53). She notes that, in addition to the three naval vessels identified by Murdo Morrison, one of which – HMS Ajax – broke down en route, three other ships joined "the armada" – the Belleisle, Amelia and Forester (Buchanan, 1996: 54). In the event, however, and despite "the large police and military presence", the situation was resolved through the interventions of two police constables

(Buchanan, 1996: 56). Of those initially arrested, only six were sent to trial at the High Court in Edinburgh (Buchanan, 1996: 58). It took the jury only half an hour to acquit them of all charges (Buchanan, 1996: 62). (With respect to the memorialization of these events in the 1990s, see Withers, 1996.)

7 No indication is given in the literature as to the relative power that women and men wielded in the tuath. With respect to clan society, Newton (2009: 158) notes that the limited sources that allow a gendered analysis recognize that what power women could wield was "informal" rather than "formal". It is nevertheless also the case that women's position would vary in relation to their social class.

8 Dwelly (1994: 422) writes of the rendition of feannagan (singular, feannag) as "lazy-bed" in English as "a southern odium on the system of farming in Gaeldom, where soil was scarce, and where bog-land could not be cultivated in any other way".

9 The most comprehensive local account is for the island of Scarp by John Maclennan (2001). The island is located just to the west of the North Harris Estate.

10 Commonties were (and are) a particular form of common property, which Callander (1998: 146) refers to as "shared private property". "[R]ights of ownership in a commonty", he writes (Callander, 1998: 145), were "based on the ownership of neighbouring (but not necessarily adjoining) land". They were a shared resource to which land owners and their tenants had rights (Callander, 1998: 145).

11 Callander (1998: 45–46) claims Scotland's distinctiveness *vis à vis* England in terms of property law on the following grounds: "Conventional feudal theory and practice was based (in England and elsewhere) on the premise that a kingdom was first and foremost a feudal entity and in that sense, the property of its king or queen. In Scotland's feudal system, this situation was radically tempered by the Crown's status as representative of the Community of the Realm, which vested that 'ownership' in the sovereignty of the people". Sovereignty had, in other words, never been ceded to a monarch (or a parliament) as in England, but remained with the people (see MacCormick, 1998) – hence the naming of the present monarch as the Queen of Scots, not of Scotland, in contrast to her naming as the Queen of England.

12 Although the Abolition of Feudal Tenure (Scotland) Act was passed in 2000, it was "brought wholly into force" with two other pieces of legislation – the Title Conditions (Scotland) Act 2003 and the Tenements (Scotland) Act 2004 – on 28 November 2004 (Steven, 2004: 1).

13 Andrew Steven (2004: 5) notes that Part 3 of the 2000 Act provides for a mechanism through which compensation can be paid to superiors "in respect of their loss of the periodical payment that is feuduty". But this, he remarks (Steven, 2004: 5), "is merely a tidying-up exercise". Under the new legislation, superiors are not allowed to "enforce" "real

burdens" – "perpetual conditions affecting the land" (Steven, 2004: 3, 5). The Act does, however, "enable them to register a notice to preserve their rights in very limited circumstances where this is considered to be legitimate" (Steven, 2004: 6). Further, limited burdens may be enforceable under the Title Conditions (Scotland) Act 2003 as "personal real burdens" (Steven, 2004: 6). One example of this obtains in the case where the superior is a conservation body, so designated on a list produced by the Scottish Ministers, and wishes to enforce a "conservation burden" aimed at protecting the natural or built heritage. In this situation, the conservation body had to register a "notice" by 26 November 2004 in order to be able to do so (Steven, 2004: 6).

14 For discussion of women's participation in the land wars, see Robertson, 1997; Mackenzie, 2010a.

15 The LRPG's first report, *Identifying the Problems*, was published in February 1998. Following extensive public consultation, *Identifying the Solutions* was published in September 1998. A third report, *Recommendations for Action*, was published in January 1999. For a discussion of these reports, see Cameron (2001). Following these reports, the government, then named the Scottish Executive, published *Land Reform: Proposals for Legislation* (Scottish Executive, 1999b).

16 The Land Fund was closed in 2006, to be replaced in May of that year by Growing Community Assets, funded by the Big Lottery. Whereas the Land Fund focused on community land acquisition, its replacement's remit was broader. The projects it supported tended to be more complex, the application process more cumbersome and the success rate lower (Gerrard, 2010: 16).

17 For discussion of Part 2 of the Land Reform Act, see Carey Miller and Combe, 2006; Combe, 2006; Wightman, 2007.

18 Regarding the geographical reach of Part 3 of the Act, see Chapter 1, endnote 5.

19 Claims that the proposed land reform legislation would contravene the European Convention on Human Rights (ECHR), specifically Article 1 Protocol 1, were dismissed by law makers on the grounds that the legislation met the "acid test" of being in "the public interest" (Combe, 2006: 210). As cited by Combe (2006: 210), "Art.1 Protocol 1 provides that, 'no-one shall be deprived of his [sic] possessions except in the public interest' but the State can 'control the use of property in accordance with the general interest'". Combe (2006: 210) writes, "From Strasbourg jurisprudence, it is clear that a non-arbitrary deprivation with proper compensation reasonably related to the property value is not a breach of Art.1. ... Accordingly, the crofting community right to buy, as part of a scheme for '[e]liminating what are conceived to be social injustices', will not be in breach of Art.1, and neither will the lesser Pt 2 of first refusal" (Combe is here quoting [1986] 8 EHRR 116, in footnote 59).

20 The Crofting Reform (Scotland) Act 1976 gave individuals the right to buy the croft house, inbye land and an apportionment of the common grazings. With consolidation of crofting law in 1993, these rights are now located in sections 12–19 of the Crofters (Scotland) Act 1993 (MacAskill, 2004: 106). For further discussion, see endnote 34.

21 For discussion of the government's position, see Combe, 2006: 217–219. Combe (2006: 218) does note that "[t]o a certain extent, the Scottish [Government's] hands were tied by the reservation of company law to Westminster, making the introduction of a new juristic personality a legislative impossibility for the Scottish Parliament", a position he contrasted with that obtaining in South Africa. Here, the Communal Property Associations Act created "a new form of juristic person" allowing for "an emphasis on registration of a community's rules rather than the form the body must take" (Combe, 2006: 218).

22 The North Assynt Estate comprises 21 132 acres, of which only 50 acres are non-croft land. At the time of purchase, about 400 people resided on the estate, one-quarter of whom were "permanent non-crofting residents" (MacAskill, 1999: 35).

23 In the case of both Galson and Pàirc, the applications to the government to purchase the land concerned the common grazings only, not the inbye land, owing to the onerous mapping requirements of Part 3 of the Land Reform Act (Land Reform Act 2003 Section 73 [5] [b] [i] and [ii]). Commenting on this requirement at the Highland Council Conference on Land Reform, Inverness, 23 March 2010, John Randall, Vice-Chair, Pàirc Trust, stated that, "the detailed mapping requirements ... go well beyond what is normally required for a private sale" (Randall, 2010: 19). To this evaluation, Simon Fraser, solicitor, added that, "This extreme level of detail would cost potentially more to produce than the cost of buying the land" (Fraser, 2010: 23).

24 Present rents range from £1.50 to £6.00 a year (interview, NHT administrator, 1 April 2011).

25 Janet Hunter (2007: 10) names the Scott family's creation of "a limited number of new holdings" in North Harris in the mid-1880s in this respect. Also see Bill Lawson's (2002: Part 5) discussion of the resettlement of the forest of Harris at this time.

26 In addition to land under crofting tenure, three small areas of land subject to agricultural leases are shown in green with crosshatching on the map.

27 These words provided the headline for the front page of the *West Highland Free Press*, 19 January 2007: 1.

28 As others, including the press, were not invited, I draw here on discussions with key local people who took the initiative forward and on Janet Hunter's (2007) book on the North Harris buyout. Also, see Mackenzie, 2006b.

29 The Chair of the steering group, Kenny Mackay, was at the time Chair of Harris Development Limited, and the person responsible on a day to day basis for progressing the initiative was Duncan MacPherson, Manager of

Harris Development Limited. This organization was established in 1994, as a company limited by guarantee, with the objective of promoting appropriate economic development on the island, as specified by the Harris Integrated Development Programme, the responsibility of Frank Rennie of Lews Castle College, Stornoway. Those involved with it gained substantial experience in applying for outside funding and confidence in negotiating with donor agencies. As a gauge of its significance, the organization has recently embarked on a £1.2 million programme of heritage projects under the Landscape Partnership Scheme.

30   Scarr Hall is chair of GSH (George S. Hall) Partners in Facilities Management, a UK-based company with a network of branches world-wide. The NHT recognized him as someone who "share[d] the same aims and objectives" as the trust (NHT News Release, October 2002).

31   The NHT (2003: 9) gives the following detail regarding the purchase price of the North Harris Estate, which in total amounted to £2 200 000, plus legal fees of £40 000: The Scottish Land Fund provided £1 569 016, the Community Land Unit £392 254, with a joint contribution to legal fees of £37 489, totalling £1 998 759. In addition to these amounts, the John Muir Trust contributed £100 000, SNH £40 000 and Comhairle nan Eilean Siar £34 000; the remainder was to be secured through public appeal (MacSween, 2003: 1; *Dè Tha Dol?* 21 February 2003).

32   Panchaud's company was finally located in Panama and, following a lengthy period of negotiations, she agreed to the sale (*Stornoway Gazette*, 9 March 2006: 5; *WHFP*, 10 March 2006: 5). With the exception of 1272 acres, the estate is under crofting tenure (Graeme Scott and Co., 2004: 3). In total, there are 22 crofts in four townships. At the time of its feasibility study, the population numbered 45 residents. In addition, five crofters who worked land on the estate lived elsewhere (Graeme Scott and Co., 2004: 13).

33   One map identifies the "Security Area", designating non-crofting land that is secured against speculative sale through a legally binding agreement with Highlands and Islands Enterprise (for 10 years) and the Scottish Land Fund (for 80 years) (interview, NHT Land Manager, 13 May 2003). In other words, except for the sale of small parcels of land, for example for housing or business units, the NHT would have to obtain the agreement of these parties before a sale were to proceed. Thus is the broader social interest in the land achieved.

34   McCrone (1997: 22) writes that the individual right to buy, introduced in the 1976 legislation, creating the category of owner-occupier, "fitted the ideological predispositions of a government keen to privatise state resources, in the only way it approved of, namely, by treating people as absolute owners". Despite the fact that the crofter could purchase the croft for a sum amounting to 15 times the annual "fair rent" – a sum that in most cases was under £300 – the provision had not proved popular (Hunter, 1991: 147). Among the reasons for the unpopularity were the means testing

for certain grants and ineligibility for other grants or loans, including those for crofter housing (Hunter, 1991: 147). Of the 17 778 registered crofts across what used to be known as the "crofting counties", now the local authorities of Eilean Siar, Highland, Orkney and Shetland, Argyll and Bute, Logie (2007: 3) writes that 3686 were owner-occupied at his time of writing. The percentage of crofts in individual ownership varies geographically. In the Western Isles, only 1 per cent of the total number of 6032 crofts falls into this category. This figure contrasts with the much higher figures for Skye, Lochalsh and Lochaber, 19 per cent of 2480 crofts being in owner-occupation, in the North West Highlands 24 per cent of 2714 crofts, in Shetland 29 per cent of 2731 crofts, in Argyll 36 per cent of the 1080 crofts, in Easter Ross, Inverness and Badenoch and Strathspey 41 per cent of the 2282 crofts and in Orkney 79 per cent of the 459 crofts (Logie, 2007: 3). Figures published in 2010 indicate that 4287 (23.9 per cent) of the 17 936 crofts were owner-occupied (Edwards, 2010: 5, cited by Busby and Macleod, 2010: 598).

35   In light of this, it is interesting to note that in March 2010, as part of the Scottish Government's consultation on crofting, Stòras Uibhist suggested that "responsibility for regulating community owned crofting estates could be delegated to those bodies" (communication with H. Francis, 18 October 2011). Their proposal, submitted to the Rural Affairs Committee of the Scottish Parliament as it deliberated amendments to the draft Crofting Reform Bill, included a call for an end to the individual right to buy and the extension of crofting tenants' rights to owner-occupiers, which would, in their view, be "a Trojan Horse for the final destruction of crofting tenure" (*WHFP*, 5 March 2010: 1). While they are far from being alone in this call – as the evidence produced for the Shucksmith Report (Committee of Inquiry 2008) demonstrated – their intervention signifies a further means through which community ownership has the potential to trouble ongoing privatization – here furthered by the individual right to buy. In the event, this recommendation was not included in the Crofting Reform Act 2010.

36   The definition of "community" in the 1997 Act, which was the Act involved in the buyout of the West Harris Estates, is not as wide as that in Part 3 of the Land Reform Act 2003 (MacAskill, 2004: 108). Quoting the Consultation Paper of 1996, which informed the 1997 legislation, MacAskill (2004: 108) writes that, "it would be for the crofters to decide whether membership of the trust should be restricted to local crofters or also include other local community representatives". Nevertheless, the West Harris Crofting Trust includes as members both crofting tenants and residents.

37   A further 173 people are registered as Friends of the NHT. These are people who live outwith the island and have no legal rights to vote on trust business.

38   Representation by geographical area is not always the case. Stòras Uibhist, for example, holds elections on a first past the post basis.

39 While focusing on crofters' rights to common grazings generally, rather than the negotiation of their rights in a situation of community land ownership, Katrina Myrvang Brown's (2007) discussion is useful in making visible the complex and sometimes contradictory claims that are made to the common grazings.

40 This information was provided by William MacDonald, Grazings Clerk, on 9 October 2008, at the time of a visit to Askernish for members of the (then) Scottish Crofting Foundation at the end of its annual gathering. Some of those opposing were absentee crofters (communication with H. Francis, Chief Executive, Stòras Uibhist, 30 August 2011). There are about 1000 crofts on the estate (communication with H. Francis, Chief Executive, Stòras Uibhist, 30 August 2011).

41 A community consultation carried out at the end of 2010 and beginning of 2011 indicated a high level (73 per cent) of support for the Askernish Golf Course Development (communication, H. Francis, Chief Executive, Stòras Uibhist, 18 October 2011).

42 Souming refers to the number and kind of livestock (cattle, sheep) to which a shareholder of the common grazings is entitled. For contemporary information on souming, see Gwyn Jones (2011: 27–30).

3
Working Nature

Introduction

Once every 18.6 years – most recently over the period 2006 to 2007 – as seen from the standing stones at Calanais, Isle of Lewis, the full moon rises from and skims the hills of North Harris and Pàirc on its "southern standstill" or "southernmost arc" (McHardy, 2006a: 8). "The significance of this event", writes archaeologist Ian McHardy, "is that the avenue and circle of standing stones at Calanais appears [sic] to have been designed to focus upon it and the ridge of Pàirc hills called, 'Cailleach na Mòinteach' [Old Woman of the Moors], which it rises from, some 4000 years ago" (McHardy, 2006a: 8). He continues,

> Watching the huge red moon being "born" from the Cailleach and skimming across her body before being framed in an avenue of standing stones was nothing short of spectacular, and would have left anyone in no doubt that this was more than some astronomical co-incidence. That our ancestors studied the heavens with such intelligence and patience cannot, to my knowledge, be demonstrated at any other such site in the British Isles. I would also not be alone in academic circles in stating that we have not yet fully understood the stones at Calanais, or indeed many other Neolithic structures to which it may provide clues (McHardy, 2006a: 8).

McHardy emphasizes in this quotation and in his precognition to the public local inquiry called to examine the proposal for a wind farm

Places of Possibility: Property, Nature and Community Land Ownership, First Edition. A. Fiona D. Mackenzie.

at Muaitheabhal on the privately owned Eisgen Estate in 2008 (McHardy, 2008) that the significance of Calainais lies not just in the stone circle – "gem of the isles" and a Scheduled Ancient Monument though it may be – but in the relationship between the stones and "the hills and skies around them, their landscape, and in particular the 'Cailleach na Mòinteach'" (McHardy, 2006a: 8). This archaeological link between the standing stones and the Cailleach is strengthened, he maintains, by the presence of a Neolithic kerbed cairn on the ridge where the Cailleach's "knee" is located (McHardy, 2008: 1).

In a subsequent publication, McHardy (2010) makes clear the significance of the association between the Cailleach and the stones. Cailleach, he writes, "is not just a word for Old Woman. ... There is a mythical 'Cailleach' of which stories are told the length and breadth of Scotland, and in closely related forms Ireland and the Isle of Man" (McHardy, 2010: 8; also see Newton, 2009). On the basis of these stories, McHardy's argument is that the Cailleach na Mòinteach "was a personification, a manifestation, of [the] ancient Earth Goddess" (2010: 15). While she is evident in many other places, this, he contends, is "the only place where she can be seen to give birth to the Moon" and, as such, the Cailleach na Mòinteach may have been especially revered, even famous" (McHardy, 2010: 15).[1]

The point, with respect to the proposal for the Muaitheabhal Wind Farm, for McHardy (2006a: 8), was that, with plans for six wind turbines to be built on the Cailleach na Mòinteach ridge, visible in daytime and silhouetted against a night sky, the event in which he had participated in June 2006 would be compromised. He had no wish to "deny the benefits of technology to the island", he said, but asked whether such benefits could be achieved "without blighting such a valuable landscape" (McHardy, 2006a: 8).

I relate this story at the beginning of a chapter whose objective is to explore how "nature" is "worked" when land is brought into community ownership in order to identify three interrelated areas of conceptualization with which I engage. First, the story of the Cailleach disrupts a norming of nature cast as ontologically separate from the social. Wrapped in mystery though the meanings of stones and hills may be (Gordon, 1995; Murray et al., 1995; Ashmore, 2002: 38–44; Burgess, 2008: 71–72; Curtis and Curtis, 2008; McHardy, 2010), there is no question that they are part of a land that has been worked for millennia – astronomically, spiritually, socially, cultur- ally, economically and politically – and where it is impossible to draw dis- tinct lines that separate "the natural" from "the social". Together with recent palaeoecological research (Gregory et al., 2005; Edwards et al., 2005; also see Burgess, 2008), the story points to a land that has been and is, at every turn, co-produced through the interrelationship between the social and the natural.[2] Second, the story – or the timing of its public telling – is a reminder that this co-production of the social and the natural is bound up with politics, in this case of the wind and the wild. In both cases, nature is contested through the assertion of rights to the land, to property. With respect to the wind, nature is performed in the story in the interest of those opposing the industrialization of a landscape by a large-scale wind farm whose benefits would overwhelmingly be reaped by the individual owner of the land (Chapter 4). Opposition centres, in other words, on the commodification of the wind in the private interest. With respect to the wild, ironic though it may be that the land to which the Cailleach belongs has so clearly been occupied since the Mesolithic and actively worked since the Neolithic, the land now, and at the time of the public local inquiry, lies within the search area for "wild" land by Scottish Natural Heritage (SNH), the statutory conservation authority (SNH, 2002). It is no coincidence that the growing interest of SNH and non-governmental organizations such as the National Trust for Scotland (2002) and the John Muir Trust (2004) in finding more robust ways to conserve wild places than currently exist is occurring at precisely the same time as the proliferation of proposals to build wind farms and erect giant pylons to take their electricity to distant markets. The windiest parts of Scotland are also the wildest and the conflict between the different sets of interests can be acute.[3] Third, the story illustrates how in conceptualizing place – Calanais – nature is not a "stable backdrop", an "eternal" (Massey in Massey et al., 2009: 412), but is implicated at every moment in the continuous re-creation of people's subject positions vis à vis place. In the narrative, the Cailleach – or the materiality of the hills through which she is brought into being – becomes one of the ways through which place is performed, its meanings changing in tandem with the continual reproduction of nature through time.

In this chapter, I draw on these ideas from the theorization of social nature to explore the ways in which nature is reworked when land becomes community property and how, in turn, the reconfiguration of nature leads to the renegotiation of people's subject position with respect to the land. My aim is to trace the political possibilities of a nature that is now "undone" through the disturbance of centuries-old processes of enclosure and privatization. I am interested in exploring how a troubling of the binary private/public as communities take over ownership and management of the land disturbs the dualism nature/culture in so far as it concerns practices of conservation and, specifically, the re-creation of "the wild".

Map 3.1 Sites of Special Scientific Interest (SSSIs) and National Scenic Area (NSA) in the Outer Hebrides. © Ashworth Maps and Interpretation Ltd 2011. SSSI and NSA boundaries © Crown copyright and database right. All rights reserved. Licence number 100035655.

I focus on three current initiatives of the North Harris Trust (NHT) to trace how this community land owning body works nature/the wild in its pursuit of a more sustainable future. As identified in one of its stated aims, "nature" is central to the reconfiguration of the estate. The NHT's aim is

> To achieve the regeneration and development of the North Harris community by managing the North Harris Estate as an area of outstandingly wild and beautiful land, through local participation and working with other partners where appropriate, in order to meet the needs, hopes and aspirations of the local community and for the benefit of the wider public.

I consider how the work of the NHT in planting "native" woodland, in the management of environmentally protected areas, and in a proposal for a national park for the Isle of Harris, contributes to this aim by "undoing" a norm of nature as distinct from social process and producing a "wild" that is a-working. Through the NHT's work, the land is shown to be a palimpsest, the different layerings suggesting that the wild is just one more way through which the land is conjured and people rebind themselves to place. This new working counters colonizing and class-based configurations of conservation and opens up the meanings of nature and the wild to possibilities that are more in line with the social and political objectives of community land ownership.

Extensive parts of the Outer Hebrides are subject to multiple and frequently overlapping protective environmental designations whose provenance lies outwith the islands. The whole of the North Harris Estate lies within the South Lewis, Harris, and North Uist National Scenic Area, a Scottish designation that recognizes "[n]ationally important areas of outstanding natural beauty" (Warren, 2009: 221). 32 436 acres (13 132 ha) of the estate are designated as a Special Area of Conservation under the EU Habitats Directive for protecting biodiversity and a Special Protection Area under the EU Birds Directive (SNH, 2007: 5). Both the Special Area of Conservation and the Special Protection Area lie within the 32 520 acres (13 166 ha) Site of Special Scientific Interest, a UK designation (SNH, 2007: 5; Map 3.1).

These designations, together with the casting of the estate in terms of "blood sports" (Lorimer, 2000; also see MacMillan *et al.*, 2010), the purview of the land owning class, have, until community ownership of the land, worked as a political technology (Foucault, 1979), removing from the visual field other, more socially just, ways of seeing. They normalized a particular way of viewing the land – as empty of people – while simultaneously masking the workings of power through particular practices of nature – as wild – dependent on separating the social from the natural (Mackenzie, 2006b: 389). Paralleling the rainforest politics of Clayoquot Sound on Canada's Vancouver Island, nature became bound up in "itineraries of silencing" (Braun, 2002: 31), erasing other claims to a land that has been worked for thousands of years by the ancestors of many of those present as crofters on the land. Unlike the politics of Clayoquot Sound, which have to do with the construction of race – a "politics of indigeneity" (Braun, 2002) – nature's "complicity" in removing people from the land in the Highlands and Islands turns on both the exercise of a colonizing conservationist optic from outwith the island and of class privilege within it. The NHT opposed this optic and this privilege when it bought the estate – without the castle of Amhuinnsuidhe – in 2003.

Native woodlands

The Millennium Walk, located in Gleann Bhioigadail near the Scaladale Centre where the celebrations for the NHT's buyout of the estate took place in 2003, stretches through 125 ha of native woodland to a knoll overlooking Loch Seaforth (Photograph 3.1). An Cliseam, the highest hill in the Outer Hebrides, rises to the west behind it. Wheelchair accessible, it is the return journey of 2000 m that earns the walk the status of being a millennium project. It was completed in 2001. The inspiration of Murdo Morrison of the Aird a'Mhulaidh grazings committee, the path passes through the first sizable planting of native woodland in North Harris on land that is part of the common grazings. The woodland consists of a mixture of rowan, oak, birch, holly, juniper, willow and alder, 210 000 young trees in total, protected from deer and sheep by a 6000 m long perimeter fence. With paid and volunteer labour, the work involved had been intensive. Drainage ditches had to be dug, soil from which was heaped in mounds in which the trees were planted, and fertilizer – delivered by helicopter, between trips to fish farms – had to be applied.[4] While still small, the trees stand in visual contrast to the unplanted land surrounding the woodland.

The NHT has added to the acreage under native woodland, amounting to 12 ha in Gleann Langadal and 6 ha in Gleann Mhiabhaig. The woodland in Gleann Langadal, as in Aird a'Mhulaidh, was planted for its amenity value, to support biodiversity and to provide employment. In addition to the species planted at Aird a'Mhulaidh, oak and Scots pine seedlings have been introduced. Although these trees are no longer found

Photograph 3.1   The Aird a'Mhulaidh Woodland and walkway.
Source: author.

in the remnants of native woodland in the islands, evidence from pollen analysis indicates their presence in the past. Unlike the millennium woodland, this woodland is planted on land that used to be the "heartland" of what was previously a sporting estate, not on land under crofting tenure (interview, NHT Land Manager, 28 March 2011).[5] The riparian planting in Gleann Mhiabhaig takes place up to 5 m on either side of the river, land owned by the Amhuinnsuidhe Castle Estate, and on adjacent land a short distance beyond owned by the NHT.

Its primary objective is to protect the land adjacent to the river and thereby provide a healthier ecosystem for the highly valued salmon. Although Amhuinnsuidhe Castle Estate owns the rights to the salmon rivers and 5 m of land on either side of the rivers, except for the case of the Gobhaig system in the west, the NHT is responsible for the environmental management of the watershed areas and river banks (NHT, 2007: 15). The riparian plantings reduce siltation in rivers, increase the nutrients available for fish from leaf-litter and invertebrates and provide shade (Parrott and MacKenzie, 2000: 4). On certain rivers, the plantings support a healthy environment for the globally rare freshwater pearl mussel, part of whose life cycle, as glochidia, is spent nestled in the gills of the salmon (Parrott and MacKenzie, 2000: 4; Hastie and Young, 2003; Sime, 2003; Skinner et al., 2003).

In the past, Murdo Morrison obtained seedlings from Christie's of Fochabers, in Moray, for the plantings in Aird a'Mhulaidh, while the NHT sourced its seedlings from Alba Trees, Gladsmuir, East Lothian, in both cases using seeds of West Highland provenance. However, the Western Isles Woodland Officer's documentation of significant woodland biodiversity in two geos that descend steeply and narrowly down to Loch Seaforth in 2004 on Rhenigeadal common grazings land, and the more recent identification of other sites in Harris and South Lochs, Isle of Lewis, by the Western Isles Native Woodland Restoration Survey (SNH, 2008b), have resulted in plans to source the seeds more locally for future plantings.[6] Alba Trees has agreed both to grow seedlings from the NHT's collection of seeds in 2010 for use in 2012 and, if the NHT wishes, to assist in establishing a tree nursery on the North Harris Estate. Together with the sale of bedding plants, such a nursery could provide at least half time employment for one person (interview, NHT Land Manager, 28 March 2011).

In planting native woodland, the NHT is in one sense part of a Scotland-wide movement, which, since the 1990s, has led to a reversal of the millennia-long trend of decline and fragmentation of native woodland, itself due both to climate change and to human action (Tipping, 2005; Smout et al., 2005).[7] Supported by national legislation on biodiversity, The Nature Conservation (Scotland) Act 2004, and by the Scottish Biodiversity Strategy It's In Your Hands (Scottish Executive, 2004), local biodiversity action plans,[8] The Crofter Forestry (Scotland) Act 1991, the Forestry Commission Scotland's Scottish Forestry Strategy (2006)[9] and the efforts of non-governmental organizations (Reforesting Scotland being a prime example), the extent of land under native woodland has increased substantially.[10] Scotland-wide, between 2000 and 2006, 9500 ha of native woodland were either planted or restored, leading Warren (2009: 291) to remark that these woodlands were no longer "a fringe concern but ... have now taken their place at the heart of forest policy"(for example, see Forestry Commission Scotland, 2006).

On the basis of this rapid expansion of land planted to native woodland and their analysis of the Forest of Spey, an area of native woodland managed by the Cairngorms Partnership Board whose members comprise individuals from both the public and private sectors, Paul Robbins and Alistair Fraser (2003: 114) contend that the planting of native woodland in Scotland, of primary value to the tourist in search of "wilderness", may be explained in terms of an "ecological fix" at a particular point – in time and space – of capitalist accumulation. These are "forests of consumption" – "the spatial product of schizophrenic capitalist ecology" – whose expansion takes place simultaneously with that of the monocultural "forests of production" (Robbins and Fraser, 2003: 112). They are, the authors suggest, "different layers of capitalist development [which] vie with one another on the landscape" (Robbins and Fraser, 2003: 114).

I do not deny that, materially, native woodlands may be "specifically intended for consumption as wild space" – whether by tourists on whom the gaze of Robbins and Fraser (2003: 108) is fixed or sport fishers who stand to benefit from any increase in salmon stock attributable to riparian planting. In either case, the woodlands constitute a means through which capital may be accumulated, as proposed in Escobar's (1995) postmodern form of ecological capital to which the authors refer. Nor, in the context of debates about "native" and "alien" species (see Warren, 2007),[11] do I disagree with their claim that the "authenticity" of native woodland as "wild" is metaphorically bound to a particular invention of the past – an Eden, a place of origin that for some environmentalists immediately pre-dates the depredations of sheep and then deer (Robbins and Fraser, 2003: 110) following the Clearances. To the extent that this reconstruction of the land relies on the regeneration of the iconic Scots pine, it may also be allied discursively to the "collectivist striving" of an "emergent nationalism" (Robbins and Fraser, 2003: 110). Drawing on the work of Fraser MacDonald (1998), Robbins and Fraser (2003: 111) attest that SNH has been at the forefront of initiatives that recall a collective claim to "our natural heritage", strategically using such symbolic devices as "native" species and, specifically, the Scots pine to assert and legitimate a particular political position, *contra* the frequently different interests of crofters and the owners of private estates.

However, I would suggest, the argument of Robbins and Fraser (2003) falls short of capturing the complexity of the material and metaphorical meanings of the resurgence of native woodland in Scotland. This results from their neglect of community plantings, which do not fit easily within the proffered binary of a schizophrenic forestry, specifically where such plantings occur on land in community ownership. However varied in form, community woodlands are central to what Warren (2009: 295) calls "the native woodland movement". Some, such as the Laggan Forest Partnership, Abriachan Community Forest and the Sunart Oakwood Project, which he cites, have achieved considerable political visibility (Warren, 2009: 88). Their objectives are complex and may indeed include a wilding of the land to increase its amenity value – for local people as well as for tourists – frequently authorized through discourses of biodiversity, the mitigation of the effects of climate change and, more generally, sustainability. However, some community woodlands also have to do with establishing rights to land – to a new configuration of property and an ontologically incompatible working of nature informed by a community-centric, rather than privatizing, ethic. About these matters Robbins and Fraser are silent.

As is the case with respect to the politics and policies surrounding native species, so "community" is a central political and policy ingredient for the Scottish Government (2009), for local authorities (for example, Comhairle nan Eilean Siar, 2003), and for enterprise companies (for example, HIE's Strengthening Communities programme; www.hie. co.uk). Community woodlands themselves are supported by a raft of measures including *The Scottish Forestry Strategy* (Forestry Commission Scotland, 2006), SNH's promotion of their Native Woodland Model (2004) and, at the turn of the century, the Millennium Forest for Scotland Project (Mackenzie, 2002). Community management and ownership of forests was furthered by the National Forest Land Scheme, introduced in 2005 by the Forestry Commission Scotland. Under this scheme, communities may buy or lease land held by the Forestry Commission, proposing such an arrangement before land has been placed on the market, in some instances at discounted rates (Warren, 2009: 89).

It could be argued, in line with McCarthy's (2005b) analysis of community forestry in the USA, that policies supporting community woodland in Scotland are congruent with "roll-out" neoliberalism (Peck and Tickell, 2002), which, since the 1990s, has had in its sights, among other measures, local empowerment of civil society – "community" – and an emphasis on "best practices" rather than the pursuit of "rigid ideological principles" (McCarthy, 2005b: 998). However, I would counter this argument by proposing that, where the planting of native woodland is tied to collective ownership of the land – the re-establishment of a common right and the reworking of a nature of which people are ontologically part – the plantings suggest an "oppositional" rather than "supplementary" relationship to this form of neoliberalism if McCarthy's (2005b: 1010) distinction is to be observed. Crofting communities that have now taken over the ownership of private estates are not simply disrupting one of a number of "core neoliberal presumptions", criticism of which McCarthy (2005b: 1010) takes as an indication of opposition, but have reversed that key marker of neoliberalism – the enclosure and privatization of property. In turn, this reversal of the norming of property in the private interest opens up the possibility of practising nature in a different way, disturbing the binary culture/nature on which neoliberalization relies.

Visually, the plantings of native woodland in North Harris disturb a particular way of seeing the wild – as an "original" landscape, a place "empty" of people and trees and to be preserved as such in the class-based interest of deer stalking and maintaining the value of the land as a sporting estate in global property markets. The action of the spade digging a hole in which to plant a seedling disrupts this visuality. Its exposure of the roots of an historically distant woodland, preserved in the acidity of the peat, troubles the "casting" of the tree-less land as wild, as in some sense subject only to the processes of a non-human nature (Mackenzie, 2006b: 390). Instead, the act of digging reveals a layered land where human practice that extends as far back as the Mesolithic is interwoven with a non-human history where, since the beginning of the present interglacial, climate change has at times favoured and at times worked against the growth of trees. As Richard Tipping's (2005) research attests, different species of trees arrived in Scotland at different times from different places and performed differently in different places in Scotland during warmer and colder, drier and wetter, climatic periods. There is no original state to which to revert, no native purity through which to authenticate the present.

The present wilding of the land with the planting of native woodland is thus but one more layering of a land, another means through which human and non-human nature inter-work. People decide what counts as native. Their labour – paid and volunteer – digs the holes and drainage ditches, plants the seedlings and positions a fence such that deer – ironically, another icon of the wild – are kept out. However, in contrast to the previous working of the land as wild – in the interest of the landed class or for the purposes of conservation – this politics of nature is bound to a collective ethic informed by community ownership of the land. The planting of woodland, particularly where it occurs in the "heartland" of the estate, is thus part of a counternarrative, which, while borrowing from a neoliberal framing of forestry in Scotland, charts a course that conjures a more socially just nature from the wilding of the land than that which preceded it.

The seeds of native trees, subject though they may be to debates about their origin, are complicit in this counternarrative. Enclosed for hundreds of years in the crevices of the land by the depredations of sheep and deer, they are now conscripted in the exercise once again of a collective right to land and to woods. On the one hand, the seeds connect genetically with a past where a different ontology of nature to that which prevails today was in evidence – a nature not construed as external to the social, but of which people were ineluctably part (see Hunter, 1995a). Trees mediated this socio-natural terrain. One source of evidence is linguistic. Each letter of the ancient 18-letter Gaelic alphabet (*A'Chraobh Ogam*, literally "tree alphabet"), corresponded with a species of (native) tree. As examples, the rowan, in ancient Gaelic *luis* (in modern Scots Gaelic *caorann*), represented the letter L; the birch, in ancient Gaelic *beith(e)* (now *beithe*), the letter B (Nic Bheatha/Beith, 2000: 19).[12] In addition, the same words in ancient Gaelic were used to identify parts of the human body and parts of trees: branches and roots correspond to arms and legs, sap to blood, leaves to hair (Newton, 2009: 239, 290). To broaden the scale, the tree provided the language for the organization of society – of the family and the wider political unit (Newton, 2009: 239–240).

Poetry provides another source of evidence, demonstrating how a person's identity was, metaphorically, bound to trees. Sileas na Ceapaich's elegy to Alastair MacDonald of Glengarry written *circa* 1721 illustrates the point, as shown in the following verse. She writes

> *Bu tu 'n t-iubhar à a' choillidh*
> *Bu tu 'n darach daingeann làidir*
> *Bu tu 'n cuileann, bu tu 'n droigheann*
> *Bu tu 'n t-abhall molach blàthmhor.*

(In translation: "You were the yew in the forest; you were the strong and steadfast oak; you were the holly, you were the blackthorn; you were the rough, blossoming apple tree") (cited by Newton, 2009: 291). Archaeological and place name evidence points towards the centrality of trees in Gaelic spirituality. Symbolically, it was the "sacred tree", "arguably the most enduring and evocative symbol in Gaelic culture", that, as "*axis mundi*" (pillar of the world), connected "heaven" and "earth", writes Newton (2009: 237). The persisting practice in the Highlands and Islands of planting a rowan beside a house in acknowledgement of its perceived protective properties is one indicator of the more general reverence accorded trees in the past (Mackenzie, 2002: 549).

On the other hand, at the same time as the seeds reaffirm an historically deep common right to the land, they mobilize contemporary global discourses of biodiversity, of measures to mitigate the effects of climate change, and of sustainability. While these discourses may well intersect with neoliberal agendas – of the expansion of native woodland and of community – they simultaneously open nature/the wild to new meanings. The day to day working of a collective ethic – the choice and collection of seeds and their germination, the digging of holes for the seedlings and trenches for drainage, the planting of seedlings and their protection through fencing – subverts boundaries between the natural and the social. This is a wild in whose creation people's labour is central and through whose creation people reposition themselves *vis à vis* nature. As Jill Casid (2005: 197) reveals in her research into "counterdiscourses of landscape" in the Caribbean, the seeds are implicated in a process of resubjectivation. They gesture towards a new politics of nature.

Managing protected areas

Since its establishment through the Natural Heritage (Scotland) Act 1991, SNH has repeatedly had to ward off accusations of "bureaucratic, high-handed interference" from farmers and crofters who work the land (Warren, 2009: 223) (albeit from ontologically different subject positions *vis à vis* nature [MacDonald, 1998; Toogood, 2003]) on the one hand and of "ineffectiveness and compromise" from conservationists on the other (Warren, 2009:223). This results, as Charles Warren (2009: 222) explains, from the merger in the new statutory body of two organizations, the Countryside Commission for Scotland – with an interest in landscape and people – and the Scottish section of the UK-wide Nature Conservancy Council – with an interest in "nature conservation". As stated in the Act of 1991, the organization's remit is:

> To secure the conservation and enhancement of, and to foster understanding and facilitate the enjoyment of, the natural heritage of Scotland. ... [The natural heritage includes] the flora and fauna of Scotland, its geological and physiographic features, its natural beauty and amenity (cited by Warren, 2009: 222).

Sustainability was to be the new organization's "guiding principle" (Warren, 2009: 222). While in theory the merger brought an official end to the Great Divide, namely, the perception that humans are separate from nature, in practice fulfilling its mandate continues to be a challenge. SNH frequently finds itself the centre of controversy, as I show in Chapter 4.

As one point of entanglement, what counts as the "wild" or "wild land" continues to be the source of debate for SNH (for example, see McMorran *et al.*, 2006; Carver *et al.*, 2008; Carver, 2009; Market Research Partners, 2008). SNH takes its cue from *National Planning Policy Guideline (NPPG) 14: Natural Heritage* (Scottish Executive, 1999a: 16), the first occasion where the wild is defined in the official lexicon. In this document, wild land is taken as "uninhabited and often relatively inaccessible countryside where the influence of human activity on the character and quality of the environment has been minimal" (Scottish Executive, 1999a: 16). Emphasizing its significance, *NPPG 14* states further that,

> Some of Scotland's remoter mountain and coastal areas possess an elemental quality from which many people derive psychological and spiritual benefits. Such areas are very sensitive to any form of development or intrusive human activity and planning authorities should take great care to safeguard their wild land character (Scottish Executive, 1999a: 4).

Following suit, SNH uses the terms "wild land" and "wildness",[13] rather than "wilderness", in its key policy statement, *Wildness in Scotland's Countryside* (2002), on the grounds that the term wilderness "implies a more pristine setting than we can ever experience in our countryside, where most wild land shows some effects from past human use" (SNH, 2002: 2). In the document, SNH recognizes that wild land has been and is being worked. If there is visible archaeological evidence of past human occupation, then "this is usually a light imprint on the land", it is suggested (SNH, 2002: 3). "The effects of past human uses", writes SNH (2002: 3),

> – where recognisable to the visitor – can be a reminder of a long history of people's past uses of these settings; these effects are rarely of a scale sufficient to be seen as an encroachment on wildness, and they can be seen as a reinforcement to the richness of the landscape history of the area.

And with respect to the present, SNH notes that, while land "most valued for its wild qualities" has had "no recent, permanent occupation", "virtually all this land is of some use for its owner – say, for grazing or often for hunting" (SNH, 2002: 3).[14] However, alongside such statements as these that suggest a contingent reading of the wild sit others that imply subscription to an original state, to the wild as "a pristine exterior, the touchstone of an original nature" (Whatmore, 2002: 9). None of the land is "truly wild" (SNH, 2002: 2) or "wholly natural", writes SNH (SNH, 2002: 4). Alongside a recognition of a wild that is bound to human working is a nature for which the wild provides "a sanctuary" – a retreat "for a nature in retreat" (SNH, 2002: 4).

In view of the perceived and growing threat to wild land from a number of sources, including large-scale wind farms, SNH advocates "a number of complementary approaches" in its policy statement (SNH, 2002: 9). These include increasing the effectiveness of the National Scenic Area designation within which areas "many (but not all) of the prime wild areas" (SNH, 2002: 9) are located,[15] working with those responsible for implementing planning guidelines such as *NPPG 6, Renewable Energy*, and establishing "collaborative management approaches" with the boards of national parks and with local communities and other land users (SNH, 2002: 9–10). It is this collaboration with local communities through what is known as a Section 15 Agreement, I argue in this part of the chapter, that allows SNH to shift from a "colonial" model to a more "*social* model of conservation" (Toogood, 2003: 163, emphasis his). SNH's adoption of a more community-centric optic suggests a troubling of the binary nature/ culture. No longer are people considered as separate from the interests of conservation of plants and animals – non-human nature – but as part of the working of these designations. The model, Toogood (2003: 163) states, "better relates to the self-definition of Highland culture and to political aspirations for community control and management". It is congruent with a discourse to which (many) crofters and other Highlanders subscribe – "that humans and nature are an indivisible whole" (Toogood, 2003: 161) – and which sustains their historical claim to sustainable stewardship of the land (Hunter, 1991). As one crofter said to me, "We were guardians [of the land] before Scottish Natural Heritage" (interview, July 2004).

My focus is the five-year management agreement negotiated between SNH and the NHT, signed in May 2007. It follows what is frequently regarded as the successful Lewis Peatlands Management Scheme, introduced in 2000, now amalgamated with the Mòinteach Scadabhaigh Management Scheme to form the new Sgeama Stiùiridh Mòinteach Nan Eilean Siar/Western Isles Peatlands Management Scheme. With its payments for the implementation of "best practice" in agricultural and sporting land use and in peat cutting in an area designated as a Special Protection Area and a Special Area of Conservation (SNH Information Sheet, *Sgeama Stiùiridh Mòinteach Nan Eilean Siar*, no date), SNH has combatted criticism about the sterilizing effects of environmental designations. In this case, those eligible are common grazings committees, owner-occupiers of crofts, crofters and agricultural tenants and landlords (SNH Information Sheet, *Sgeama Stiùiridh Mòinteach Nan Eilean Siar*, no date). In the case of North Harris, the agreement has been negotiated with the land owning community body, the NHT. It is this agreement that, I show, has opened up the meanings of nature and the wild such that the environmental designations now provide a place where people's collective bond to the land is renegotiated and where previously existing social boundaries are reworked.

The Section 15 Agreement signed by SNH and the NHT aims "to support" the NHT "in the long term management of North Harris SSSI [Site of Special Scientific Interest] and particularly in relation to its aims for long term sustainable conservation management of the land" (SNH, 2007: 1). As I have already indicated, the SSSI designation occupies 13 166 hectares of the western part of the estate, an area within which lie a Special Area of Conservation and a Special Protection Area. The entire area lies within the South Lewis, Harris, and North Uist National Scenic Area. About 2014 ha of this area are subject to crofting tenure (SNH, 2007: 5). For SNH, the area has also been the subject of its search for wild land.

A primary management objective of the Agreement is "to [halt] the decline in and then to [enhance] the condition of the 9 SAC qualifying features presently classed as being in unfavourable condition": acidic scree, alpine and subalpine heaths, blanket bog, depressions on peat substrates, wet heathland with cross-leaved heath, freshwater pearl mussel, montane acidic grasslands, salmon and dry heaths (SNH, 2007: 8). Those species considered to be in a healthy condition, such as the golden eagle, a qualifying species for the Special Protection Area and one of the conservation interests of the SSSI, are not included in this list. Other management objectives focus on both ecology (including support for biodiversity and appropriate land management) and human use of the land (including recreation and protection of "the landscape interest of the Land") (SNH, 2007: 8–9). Over a five year period, in return for the NHT's work, SNH agrees to pay a total sum of £70 675 (SNH, 2007: 4). While most of these funds fall into the category of "General Management", specific sums are allocated for deer counts and culls and habitat monitoring. The money has been primarily used to pay part of the salary of the NHT Land Manager, who has overall responsibility for ensuring that the agreement is fulfilled.

As part of the Agreement, the NHT was required to produce a land management plan within one year of signing. This plan was to "adopt a holistic approach to the Land", which included policies about the management of land under crofting tenure, sport, access, recreation and education, economic development and monitoring and evaluation (SNH, 2007: 9). In recognition of the impact of deer on habitat, SNH also required the NHT to complete a deer management plan to be agreed to by SNH and approved by the Deer Commission for Scotland (SNH, 2007; see NHT, 2009a). The result was a *Land Management Plan* (NHT, 2007) which is comprehensive in scope. It identifies specific ecological concerns and how they will be addressed, including identifying specific issues where more information is needed. It makes evident the NHT's awareness of the global significance of some of the species and habitats for which it will have responsibility. It assesses issues relating to people, society, culture and economy, recognizing the interweaving of the ecological with the social.

The stakes for conservation are high. To cite one example, Scotland is now considered to be one of the few remaining strongholds for the globally rare freshwater pearl mussel (*Margaritifera margaritifera*), holding about half of the known "viable" populations (Hastie, 2004: 2; also see Cosgrove *et al.*, 2000). Within the UK, the mussel is protected under the Wildlife and Countryside Act 1981 and is identified as a "Priority Species" in the Biodiversity Action Plan (Hastie, 2004: 2). In the EU, it is listed in Annexes II and V of the Habitats Directive (Hastie, 2004: 2). It is also cited in Appendix III of the Bern Convention (Hastie, 2004: 2). It is classified as "Endangered" in the IUCN 1996 Red Data Book (Seller, 2006: 1). It is the primary qualifying interest for the North Harris Special Area of Conservation, the Atlantic salmon (*Salmo salar*) and the otter (*Lutra lutra*) being also named as qualifying features (Seller, 2006: 12). With respect to the North Harris site, P. J. Cosgrove and J. E. Farquhar (1999: 15) go so far as to state that one of the rivers in North Harris is "one of the top 15 *M. Margaritifera* rivers in Scotland". They add that the river "is a very important site in global terms and provides the species with a refuge population unparalleled in the Western Isles" (Cosgrove and Farquhar, 1999: 15).

While healthy, functional, beds of freshwater pearl mussels have been evident until recently in one of the two rivers with sizable populations (Cosgrove and Farquhar, 1999) – i.e., there is a well-balanced age structure in the mussel beds, with active recruitment of juveniles – recent losses have changed this picture (Hastie, 2004).[16] Lee Hastie (2004) attributes the losses primarily to illegal freshwater pearl fishing[17]; Jonathan Seller (2006: 7) considers that the greatest threat in the main river comes from its "highly dynamic" nature. There are "regular and large fluctuations in discharge resulting in an ever changing morphology" (Seller, 2006: 7). Stream braiding can result in the isolation of parts of the rivers from the main water course, with the result that pearl mussels are put at risk (Seller, 2006: 7). Both Hastie (2004: 10) and Seller (2006: 8) recommend the close monitoring of the sites, something that is more effectively managed when responsibility is local – whether by river watchers employed by the Amhuinnsuidhe Castle Estate or, in the case of the river owned by the NHT, by members of the community and the NHT Ranger.

My point here is that, subject to the condition required by the management agreement that no unauthorized changes are instigated without SNH's prior written consent, the NHT now has responsibility for the estate's "natural heritage" – of global as well as local significance. Recalling crofters' historically legitimated ethic of environmental stewardship, itself predicated on the claim to an integral bond to the land/nature, the NHT is now re-claiming the right to manage the land in the interest of the whole community – crofters and non-crofters. This is not a right handed down intact from the past, but an historically sanctioned right that is now negotiated on an ongoing basis with other social actors, specifically, in this case, SNH (see Toogood, 2003: 166). This process of negotiation undercuts a colonial model of conservation, calling into question the norm of nature – as separate from the social – through which it is authorized. At the same time that negotiation of the management agreement opens up the space for the repositioning of people collectively *vis à vis* nature and the wild, so does the disturbance of this norm of nature prompt the reworking of social boundaries. Deer, once the means through which class privilege was exercised when North Harris was run as a sporting estate, now provide the NHT with the material and discursive means through which previously rigorously enforced social boundaries are first loosened and then upturned.

Among the most "genetically pure" herds in Scotland, numbering about 1110 deer in 2010, the deer are now subject to the North Harris Deer Management Plan 2009–2012 (NHT, 2009a), a requirement of the management agreement signed with SNH.[18] The plan's objective is to: "[d]evelop a well-managed deer herd grazing recovering habitats, supported by significantly increased community management capacity"

(NHT, 2009a: 1). The plan outlines how culling will be carried out on an annual basis in order to ensure that stocking densities (which are also affected by the number of sheep grazing on the land) are such that the vegetation communities for which a large part of the North Harris Estate was designated a Special Area of Conservation recover from overgrazing. It is through the arrangements made for culling the deer that social boundaries are disturbed.

First, faced with an immediate need to cull the deer to prevent further ecological damage, the NHT signed a 20 year lease for stalking rights in the western part to Amhuinnsuidhe Castle Estate and, to the east of a line from Bun Abhainn Eadarra to Loch Langabhat, following a previous informal practice, on an annual basis to Ath Linne Estate.[19] As a "blood sport" (Lorimer, 2000), deer had, in the past, provided the means of contributing to the finances of an estate whose land was now in community control. The new owner of the Amhuinnsuidhe Castle Estate, Ian Scarr Hall, having expressed an interest in working with the community trust at the time of purchase, as well as in sustaining employment (stalkers), saw this as a way to support his estate financially. Through this arrangement, class boundaries were loosened. Deer stalking remained the privilege of the moneyed elite and nature a commodity, the marketing of which served the interests of two privately owned estates. However, as owner of the land, the NHT received income from the lease of stalking rights, and thus a proportion of the income was directed towards a collectively defined interest.

Second, the social boundaries that had previously defined the "sport" were further disturbed in 2004 with the establishment of the Harris Hind Stalking Club, a measure facilitated by the NHT, which was keen to respond to local interest in stalking.[20] Membership in the club is open to residents of Harris and South Lochs, Lewis, and costs £5.00 per annum. Members were (and are) charged a nominal fee of £10 for each hind. In 2006, the club was renamed the Harris Stalking Club after the negotiation of a lease to stalk six stags (at £175 a head) on land that had previously been part of the Loch Seaforth Estate. Following the NHT's decision to terminate stalking rights to Ath Linne Estate in 2009, the land leased to the club was extended to include all the land east of the line from Bun Abhainn Eadarra to Loch Langabhat and the club was allowed to stalk up to 16 stags. As part of a broader effort to provide supplementary feeding for golden eagles,[21] in 2010, in addition to taking hinds for home consumption (50 in 2010), the club became responsible for culling an additional number (30 in 2011) whose carcases are left on the hill for the eagles between November and February (interview, NHT Land Manager, 28 March 2011; NHT Ranger Progress Report, 15 November 2010).[22]

Since its inception, the club has been run by its members separately from the NHT. Under this arrangement, income from leasing stalking rights goes to the NHT and thus contributes to the building of a community ethic. Through the club and its relationship to the NHT, deer are re-established as property held in common, recalling people's historically legitimated right to this animal (Hunter, 1995a: 64) and making visible contemporary collective rights to the land. Nature remains a commodity, but, as the fee charged by the club is low, stalking is no longer the exclusive domain of the wealthy. The boundary between local people and those who pay large sums of money to stalk is, in other words, fractured. This counternarrative of a blood sport is furthered through the club's commitment to training its members in best practice and deer biology. Of its 25 members in 2010, 15 had achieved Deer Stalking Certificate Level 1, and further training to the second level is underway (NHT, 2009a: 4).[23] Through the training, local ecological knowledge is re-created, there is a skilling of community members, and the bond to the land is, potentially, strengthened.

Negotiating a national park

The idea to explore the possibility of creating a terrestrial national park on the Isle of Harris – its territory to be coincident with the parish of Harris rather than the North Harris Estate – was presented in a memorandum from NHT director David Cameron to the NHT board to be discussed in January 2008. As expressed more fully in a subsequent press release stating that a process of community consultation was about to begin, the case for exploring the proposal through a feasibility study was compelling. The benefits were listed as follows:

- [The feasibility study] would allow the community to decide its own future – Proposals that Harris be considered for national park status have been made previously by outside bodies. It would therefore appear to be better for the community to investigate the issue and come to its own opinion.
- A land national park could bring significant job opportunities and could create considerable economic benefits e.g. Cairngorms national park employs over 60 people directly and has a turnover of several million pounds per year.
- It could offer greater outdoor recreational opportunities and facilities for locals and visitors.

- It could give a positive return to the numerous European and UK environmental designations which are already in place over the island.
- It could raise the area's profile and encourage more visitors to come to the island (NHT, 2008).

This was not, in other words, a case of "people from outside the area [being] parachuted in to tell local folk what to do", as Cameron assured the other directors in his memorandum. Rather, the initiative would lie with the people of Harris and, as allowed under the National Parks (Scotland) Act 2000, locally elected members of the park board could form the majority of the park authority, the body responsible for ensuring that the aims of the national park aims are achieved (National Parks [Scotland] Act 2000, Section 9 [1]).[24] The Act, Cameron writes, has four aims, which are similar to those of the NHT, to which I refer below. A national park on Harris, Cameron remarked in an interview, would be "like a hat stand that you can hang lots of hats on – Gaelic, the natural environment, education, heritage. ... Such an initiative would bring young people with families" (interview, 1 May 2008). Further, as he remarked in an subsequent meeting, a national park had meaning beyond the local: "It's big stuff. ... It's not local politics anymore. ... It's a Scottish issue" (interview, 10 October 2008).

Following the NHT board's endorsement of the proposal and an indication of support from then Minister of the Environment, Michael Russell (Russell, in letter to Duncan MacPherson, NHT Land Manager, March 2008), five community consultation meetings were held throughout the island. On the basis of interest evident in these meetings, a Harris National Park Study Group comprising 15 members representing different parts of the island was established with Calum MacKay, Chair of the NHT, as chair. The group's remit was to investigate the proposal further and to prepare a brief for a feasibility study. In order to inform themselves better about the two existing national parks, the group arranged for representatives from both the Cairngorms National Park and Loch Lomond and the Trossachs National Park to visit the island and meet with community members in Tairbeart and Leverburgh. Members of the study group themselves visited the two national parks and met with members of local communities, councillors, crofters, land owners, members of the board of each park authority and park employees.

The proposal to establish a national park on Harris by people on Harris is radical, first, in the sense that it is an initiative coming from the people of Harris themselves, leadership being provided by the NHT. Although it is the case that consultation occurred in the establishment of both existing parks, most fully analysed in the case of Cairngorms National Park, in both these cases the proposal was initiated from outwith the area, by SNH (Ferguson and Forster, 2005; Thompson, 2005). A proposal from the people who would be affected by this status turns upside down the conventional – "colonial" – model of conservation. Second, the initiative is radical in that, congruent with the National Parks (Scotland) Act 2000, it troubles the norm that elsewhere frequently underpins this designation, namely the separation of the "natural" from the "social". I turn to this issue first before returning to matters of governance.

The Scottish legislation, as Warren (2009: 239) observes, was drafted in light of the recognition that substantial areas of the two proposed parks were "working landscapes with rich human histories".[25] It was also crafted on the basis of experience elsewhere. This experience included, first, a distancing from the ""wilderness" model" of the "original" national parks of the USA, which focused on "wilderness conservation" and required public ownership of the land. Second, it differed from the model adopted in England, Wales and the majority of national parks in Europe, which privileges conservation and recreation (Barker and Stockdale, 2008; Stockdale and Barker, 2009). The National Parks and Access to the Countryside Act 1949 of England and Wales, for example, has as its primary aims "the preservation and enhancement of natural beauty and the encouragement of public enjoyment" (Stockdale and Barker, 2009: 481), aims that did not shake the ontological premise on which such a designation is based. Like the legislation pertaining to England and Wales, no transfer of land into public ownership is required under the Scottish Act. However, Schedule 2 of the Act does allow for the acquisition of land "by agreement" or, subject to the stipulations of other legislation, "if authorised by the Scottish Ministers", for the compulsory purchase of any land within the park boundaries (National Parks [Scotland] Act 2000, Schedule 2, Section 5).

While the Scottish Act shares two of its aims with the legislation for national parks south of the border (identified below as (a) and (c)), its explicit commitment to "sustainability" is distinctive. As defined in the Act, the aims are:

(a)   to conserve and enhance the natural and cultural heritage of the area,
(b)   to promote sustainable use of the natural resources of the area,
(c)   to promote understanding and enjoyment (including enjoyment in the form of recreation) of the special qualities of the area by the public, and

(d) to promote sustainable economic and social development of the area's communities (National Parks [Scotland] Act 2002, Section 1).

If there is conflict among the aims, the Act states, the national park authority "must give greater weight to the aim set out in section 1(a)" (National Parks [Scotland] Act 2002, Section 9 [6]), a requirement that, Warren (2009: 239) suggests, allows critics to argue that the fourth aim is thereby demoted in importance.[26] As of the end of 2008, this clause had not been invoked in either of the existing national parks, despite the emergence of a number of controversial issues (Spencer, 2008: 7).

It may nevertheless be argued that recognition of the intertwining of the "social" and the "natural" through the aims of the legislation and the mobilization of the undergirding principle of sustainability – open though this principle may be to political conscription (see, for example, Escobar, 1995; Mackenzie, 1998b) – suggests an opening where a troubling of the two categories can occur and where more complex, mobile, and multiple configurations of each may be forged than is allowed through the 1949 legislation in England and Wales. In Steve Hinchliffe's language (2007: 99), their opening allows "a looser kind of sense", "an acknowledgment of uncertainty, a knowing of indeterminacy". The scene is, in other words, set for a more contingent working of social nature.

While unique in the UK context, Aileen Stockdale and Adam Barker (2009; also Barker and Stockdale, 2008) point out that the Scottish legislation is aligned with current thinking of the International Union for the Conservation of Nature (IUCN) for Category V areas. In contrast, for example, to Category II areas, which define national parks in terms of "American concerns of nature conservation and recreation" (Stockdale and Barker, 2009: 480), Category V areas, to which the IUCN refers as "protected landscapes" rather than "national parks", are defined as "landscapes" or "seascapes" where

the interaction of people and nature over time has produced an area of distinct character with significant aesthetic, ecological and/or cultural value, and often with high biological diversity (IUCN, 1994, cited by Stockdale and Barker, 2009: 480).

The Category V approach evolved, Stockdale and Barker (2009: 480) write, as a response both to "the emergence of sustainable development as a paradigm for protected area management and increased awareness of the 'dynamic' nature of working landscapes". "Here", they continue (Stockdale and Barker, 2009: 480), "landscapes are seen to merit special protection because they encapsulate the co-evolution of society and nature". Following from this, they conclude, "Preserving the integrity of landscape areas must therefore be based on an appreciation of the role of local communities in the processes of landscape change" (Stockdale and Barker, 2009: 481). Their maintenance is thus "dependent upon the promotion of management objectives which account for both conserva-tion and socio-economic development" (Stockdale and Barker, 2009: 479). The guidelines relating to Category V designations reflect this thinking and call, among other things, for practices of community co-management of designated areas.

This line of thought, insofar as it concerns a focus on sustainability and the need to recognize the interrelationship between people and the land, between the "social" and the "natural", was, interestingly, presaged in Scotland by the work of the Countryside Commission for Scotland, one of the precursors of SNH, in 1990 (Countryside Commission for Scotland, 1990).[27] That their proposal for national parks was rejected, as well as earlier proposals, was due to a lack of political support and a strong lobby against them by private land owners and local authorities, a matter that changed dramatically with devolution (Warren, 2009: 237). In the run-up to the referendum on devolution, as John Randall, chair of the working group that had been set up in 1995 to review all environmental designations by Michael Forsyth, then Secretary of State for Scotland for the Conservative Government, stated, Donald Dewar, the Labour Government's new Secretary of State for Scotland following the elections of 1997, was searching for ways in which Scotland could carve distinctiveness from England (interview, John Randall, 3 July 2008). The pursuit of differ-ence coincided with the working group's readiness to report, and Dewar "jumped on" the working group's consideration that "a good case" could be made for national parks (interview, John Randall, 3 July 2008). The legislation was to be distinctive on the grounds both of privileging the interworking of socio-economic and natural heritage objectives as discussed above and of the particular structures of govern-ance to which I now turn. The Act itself and the early establishment of two national parks through subsidiary legislation – Loch Lomond and the Trossachs in 2002 and Cairngorms in 2003 – were thus part and parcel of the political working of a new Scotland. In Warren's words, the legislation was "undoubtedly an expression of the post-devolution 'nation-building' agenda which sought to demonstrate the 'added value' and distinctive approach of the Scottish Parliament" (Warren, 2009: 237, drawing on Alison Rennie, 2006).

With respect to the line of argument, I draw attention to three practices of governance of the national parks through which people reposition themselves *vis à vis* the land and nature. The first concerns the powers accorded to the park authority, as legislated in the National Parks Act 2000 and as specified in the designation order specific to each park.[28] There are two alternatives. Under Section 10 (1) (a) of the Act, the national park authority may be "the planning authority for the National Park for the purposes of the planning Acts [i.e. the Act of 1997]" – with "full" planning powers – an approach adopted in Loch Lomond and the Trossachs National Park through the designation order establishing the park, an institutional practice paralleling that in England and Wales (National Parks [Scotland] Act 2000, Section 10; Stockdale and Barker, 2009: 483; see The Loch Lomond and Trossachs National Park Designation, Transitional and Consequential Provisions [Scotland] Order 2002). Alternatively, as stipulated in Section 10 (1) (b) of the Act, the national park authority may be "treated as the planning authority for the National Park, but only for such purposes of Part II (development plans) of the Town and Country Planning (Scotland) Act 1997 (c.8) as are specified in the order". As a "statutory consultee" (Barker and Stockdale, 2008: 187), it would have "call in" rather than "full" powers. This means that, following a review of all planning applications, the park authority would "call in" and "determine" those applications relevant to the park's aims (Bryden *et al.*, 2008: 34–35). This is the option taken by the Cairngorms National Park, making it unique in the UK with respect to spatial planning. In this case, "strategic planning and development control powers" rest with the local authorities (Aberdeenshire, Angus, Highland and Moray), while responsibility for the local plan lies with the park authority (National Parks [Scotland] Act 2000, Section 10 [1] [b]; Stockdale and Barker, 2009: 483; see, The Cairngorms National Park Designation, Transitional and Consequential Provisions [Scotland] Order 2003). Stockdale and Barker (2009: 483) point out that this model remains highly contentious, citing as evidence the Highland Council's position that its support for the park was conditional on the retention of planning powers by local authorities.[29]

Second, as Cameron identified in his submission to the NHT board, governance has to do with membership of the national park authority. The precise number of members – up to a maximum of 25 – is to be indicated in the designation order (National Parks [Scotland] Act 2000, Schedule 1, Section 3 [1]). Membership comprises "elected members", "nominated members" and "directly appointed members". With respect to elected members, the designation order is required to indicate:

(a)   the number of members, being at least one fifth of the total number of
      members, who are to be elected in a poll of all those who, on the day of
      the poll –
      (i)   would be entitled to vote as electors at a local government election
            in an electoral area falling wholly or partly within the National
            Park, and
      (ii) are registered in the register of local government electors at an
            address within the National Park, and
(b)   the day on which the poll at the first election under this paragraph is held
      (National Parks [Scotland] Act 2000, Schedule 1, Section 3 [2]).

There is, thus, legally, a minimum of 20 per cent but no upward
restriction on the proportion of the authority's members who are
elected to the board of the park authority, a further distinction with
respect to the legislation pertaining to England and Wales (Spencer,
2008: 8). This matter is of material significance if local democratic
accountability is to be ensured as a balance to accountability to Scottish
Ministers (see Barker and Stockdale, 2008: 187). The Scottish Ministers
are required to appoint the other members of the park authority, half of
whom are nominated by the local authorities identified in the designation
order (nominated members) and the remainder are appointed directly
by the ministers (National Parks [Scotland] Act 2000, Schedule 1,
Section 3 [3]). The Act attaches conditions regarding local consultation
and expertise to the selection of directly appointed members (National
Parks [Scotland] Act 2000, Schedule 1, Sections 5, 6). In the case of
both Loch Lomond and the Trossachs National Park and Cairngorms
National Park, five members are elected locally, 10 are nominated by the
local authorities and 10 are appointed by Scottish Ministers (Spencer,
2008: 26).

Third, with respect to matters of governance, the Act specifies that the
national park authority may negotiate a management agreement with
anyone "having an interest in land to do, or secure the doing of, whatever
the parties to the agreement consider necessary to achieve, in relation to
the National Park, the National Park aims" (National Parks [Scotland]
Act 2000, Section 15 [1]). As indicated earlier with respect to a
management agreement between the NHT and SNH, a land owning
community trust is in a strong position to negotiate such an agreement.

I turn now to the feasibility study to establish a national park on the
Isle of Harris, which the Isle of Harris National Park Study Group
commissioned on the basis of substantial local interest. The feasibility
study assesses the case for a terrestrial national park that could be made

under the terms of the National Parks Act 2000, specifically, with respect to the three conditions that such a proposal must meet:

(a) that the area is of outstanding national importance because of its natural heritage or the combination of its natural and cultural heritage,
(b) that the area has a distinctive character and a coherent identity, and
(c) that designating the area as a National Park would meet the special needs of the area and would be the best means of ensuring that the National Park aims are collectively achieved in relation to the area in a co-ordinated way (National Parks [Scotland] Act 2000, Section 2 [2]).

My focus is thus not on a proposal as such but a feasibility study that explores the different options and provides residents on the Isle of Harris with information about the extent to which a national park could contribute to the achievement of "a more sustainable future" (Bryden *et al.*, 2008: 2). In this respect, the study presents a strong case demonstrating how the three conditions identified in the legislation are met and how a national park on Harris would "fit" the four aims of such a designation as specified in the Act. Beyond this, the significance of the study lies in its potential role as "an independent assessment of options, benefits and potential disadvantages" should the proposal proceed (Bryden *et al.*, 2008: 2).

My purpose is not to assess the case for or against a national park as presented in the feasibility study. Rather, it is to identify the ways in which the feasibility study works with the categories human and non-human nature and with matters concerning governance – in light of the aims of the legislation – in order to assess the potential of a national park on the Isle of Harris to point towards a new politics of nature, one informed by a more contingent reading of nature and the wild than has frequently been the case in approaches to environmental designations. Unlike the NHT's initiatives with respect to the planting of native woodlands and a management agreement with SNH, this proposal is not restricted to land in community ownership, although, with the purchase of the West Harris Estates by the West Harris Crofting Trust in 2010 and the possibility that the island of Scalpay will soon follow suit, the percentage of the land on Harris in community ownership is growing. However, I think it is no coincidence that the initiative came from a land owning community trust, the NHT, and that leadership and logistical support are provided by that community body. Thus, while I have no grounds to suggest that community land ownership is a *sine qua non* for a contingent politics of nature with respect to a national

Table 3.1 Comparing Harris with existing UK National Parks.

Location	Area (km²)	Population[i]	Population[i] (km²)	Visitor numbers per annum	% of area under designations[ii] SAC/SPA	SSSI	NSA
Harris	500	1984	3.9	117000	31.8	36.3	100
Western Isles	3268	26400	8.6	195000	18.5/27.7	11.7	36.5
Scottish National Parks							
Cairngorms	3800	16000	4.2	1.2 million	25.0	39.0	16.2
Loch Lomond and Trossachs NP	1865	15600	8.4	3.0 million	2.9	8.6	17.2
Other UK National Parks							
The Broads	303	c5000	16.5	(iii)	23.8	24.0	N/A
New Forest	580	c34400	39.3	(iii)	48.0	(iii)	N/A
Pembrokeshire Coast	620	22542	36.4	(iii)	10.6	(iii)	N/A

i) Population figures are total population. Elected members are drawn from the electoral roll. The roll excludes those under 18 or who are not registered.
ii) Western Isles designation area percentages were obtained from the SNH website as of July 2007.
iii) No data provided in original.

Source: Bryden *et al.,* 2008: 6. Used by permission of the Isle of Harris National Park Study Group.

park, I would argue that this possibility may well be more accessible with community land ownership.

As would be expected in a study whose remit is to assess the case for the establishment of a new national park, specifically with respect to the first condition in the Act specified above, the authors of the feasibility study detail the richness of the "natural" and "cultural" heritage. In both the body of the report and in Appendices 3 (Natural Heritage Assessment) and 4 (Landscape Assessment), the study gives precise information about the various protective designations that apply to "nature" – that the entire area lies within a National Scenic Area, that 31.8 per cent of the area is designated as a Special Area of Conservation with one adjacent Special Protection Area, that five SSSIs extend over 36.3 per cent of the area, that St Kilda is designated as a World Heritage site,[30] for example (Bryden *et al.*, 2008: 17–19; Appendices 3 and 4). Table 3.1 indicates how Harris compares with other national parks in the UK in this regard. Summarizing the case for meeting Condition 1 with respect to the "natural heritage", the authors write:

> The hills, moorland, beaches and offshore islands provide a nationally recognised land and seascape. The distinctive Harris hills form the largest area of wild land in the Western Isles, and in the dramatic transition from sea to mountain summit, climate and exposure can be extreme. The relationship of the rugged mountains and the deep incised fjord lochs to the machair and peerless beaches of the west coast, when taken in conjunction with the classic cnoc and lochan landscape of the Bays area on the east coast, provides a diversity of landscape that cannot be matched in an area of comparable size anywhere in Scotland (Bryden *et al.*, 2008: 18).[31]

However, the study also recognizes the intertwining of the "natural" and the "cultural". The authors acknowledge 9000 years of human occupation of the land – from the Mesolithic, through the Neolithic and Beaker periods, to the Iron Age and Norse settlements and then to the more recent past and the activities associated with crofting and the role of Gaelic (Bryden *et al.*, 2008: 19). There is "strong interplay between the natural world, settlement and culture", the authors write; "[h]uman activity has left subtle, yet perceptible traces that contribute to the landscape and that give a strong sense of place" (Appendix 4: 11). Feannagan are singled out as a particular instance (Appendix 4: 11). Elsewhere in the report, where the writers consider the "fit" of the proposed park to the first aim identified in the Act, the low input agricultural practices associated with crofting and their contribution to sustainability – for example with respect to the corncrake and corn bunting – are highlighted (Bryden *et al.*, 2008: 24). "The natural heritage of Harris", they emphasize, "needs great care as it reflects the legacy of human activity over generations" and there is a danger that, with the recent decline in "traditional activities of stock management and small scale cropping", skills will be lost, with a resultant negative effect on that heritage (Bryden *et al.*, 2008: 23). A national park on Harris, the writers consider, "would be expected to promote objectives that would support and sustain crofting as a unique way of life, a source of good quality food, and a key feature of the island's communities" (Bryden *et al.*, 2008: 39).

Gaelic is viewed in the study as a key to understanding how people are bound to the land and how the relationship between people and nature is mediated. "The environment and Gaelic culture are inextricably related, each emanating from one another", the authors state (Appendix 4: 11). They continue:

> Greater distinction is made in Gaelic between types and categories of feature than are [sic] made in English. This is of practical use, so as to specifically name each location and make identification easy. But it also engenders an intimate "sense of place", imparting a greater landscape understanding and informing life from a Gaelic perspective (Appendix 4: 11).

The idea of a working land and of a "nature" that does not stand in originary form divorced from human activity is made evident through Gaelic as well as through the marks in and on the land itself. And it is this opening of the terms that then allows the writers to demonstrate how national park status might lead to a more sustainable future, consonant with the aims of the Act. Support for crofting has already been identified in this respect. As a further example, small-scale renewable energy installations such as a community-led wind farm would "normally be very acceptable within a National Park" and would, the study's authors argue, enable both a secure and long-lasting income stream and contribute to mitigating the effects of climate change (Bryden *et al.*, 2008: 24). Moreover, the provision of affordable housing, central to any plan to reverse population decline and stem the erosion of a deep cultural heritage centred on Gaelic, could also be furthered through a national park authority (Bryden *et al.*, 2008: 42–43). In addition to the jobs created directly through employment by the park authority, the study by Bryden *et al.* (2008: 55) estimates that 90 full-time equivalent positions could be generated if account is taken of such other indirect impacts as an increase in tourism and construction associated with the "branding" of the area as a national park.

At the same time, a repositioning of people *vis à vis* nature is proposed through the structure of the national park authority and its practices of governance. In this regard, with respect to the powers accorded the park authority, Bryden *et al.* (2008: 34–35) favour the model adopted by the Cairngorms National Park, with "call in" rather than "full" planning powers. This, the authors state, is "a proven model" in the Cairngorms National Park (Bryden *et al.*, 2008: 34). The model might employ fewer staff directly than the "full" powers planning model, but overall, it is argued, unless the "full" planning powers model received more core funding from the government than the "call in" planning powers model, the latter might generate more employment by accessing match funding to finance projects (Bryden *et al.*, 2008: 55). What they do not address in their assessment of the relative merits of each proposal is that it might well be that this model is more acceptable to Comhairle nan Eilean Siar than the model adopted by Loch Lomond and the Trossachs National Park, as the position taken by the Highland Council in the case of Cairngorms National Park demonstrated. If the "call in" planning powers model were adopted, as the park authority would not itself be directly responsible for providing a ranger service or visitor centres, it would employ 10–15 staff, paid for by core funding and finance for projects (Bryden *et al.*, 2008: 55).[32]

Of the three options Bryden *et al.* (2008: 36) consider regarding composition of the board of the national park authority (Table 3.2), the team recommends option 2 on the grounds that it is "the most balanced". Such a number in a park of this size would, they suggest, facilitate local democracy, a matter that had been affirmed as a priority both by the then Minister for the Environment, Michael Russell, and by the research team responsible for *The National Parks Strategic Review* (Spencer, 2008).[33] Introducing a debate in the Scottish Parliament on 13 March 2008 on the national parks in light of the Spencer review, while strongly endorsing their overall importance in creating a more sustainable and greener Scotland, Russell emphasized "the primacy of the local democratic element" (Scottish Parliament, 13 March 2008: col. 6973). "That element", he continued, "has served the parks well and it needs to come to the fore – indeed, if anything, it needs to be strengthened (Scottish Parliament, 13 March 2008: col. 6973). Later in his speech, strongly endorsing the actions on Harris led by the NHT, he emphasized that "communities must be seen as central to the national park process" (Scottish Parliament, 13 March 2008: col. 6976).

Table 3.2 National Parks Authority: board representation options for the full and call in models.

	Total (up to 25)	Ministerial appointees (equal number with LA)	Local authority nominees (equal number with MA)	Directly elected (at least 20%)	Comment (Harris has three community councils)
Option 1	15	3	3	9 (60%)	Three people from each community council area
Option 2	16	5	5	6 (40%)	Two people from each community council area
Option 3	15	6	6	3 (20%)	One person from each community council area

Source: Bryden *et al.*, 2008: 36. Used by permission of the Isle of Harris National Park Study Group.

For the writers of the feasibility study, people on Harris had a wealth of experience on which to draw in contributing to the management of a national park. Whether through the actions of Harris Development Ltd., Harris Voluntary Service, or the NHT, collectively, people had demonstrated resilience in the face of considerable difficulties, Bryden *et al.* (2008: 9–10) emphasize. Harris, they continue, "could make a strong case in terms of community involvement, motivation and capacity, sense of place and connectivity with culture and environment". They write:

> Harris would be a new model for a National Park, but one that is clearly within the terms and ethos of the Act. *Establishing a rationale for the applicability of National Park status to remote and economically fragile areas with strong community ethos and support would be a very significant evolution for Scottish National Parks* (Bryden *et al.*, 2008: 10; emphasis in original).

Moreover, they added later in the report, the fact that so much land is in community ownership contributes to the likelihood that sustainability of park objectives can be achieved (Bryden *et al.*, 2008: 24) (Table 3.3).

Following receipt of the feasibility study in December 2008, a postal ballot of residents on Harris was taken in February 2009. With a turnout of 71.6 per cent of those eligible casting a vote, 70 per cent (732) voted in favour of seeking national park status for the island. 311 voted against; there was one spoilt paper (e-mail communication from Duncan MacPherson, on behalf of the Isle of Harris National Park Study Group, 20 February 2009). The group recommended support for a national park on the grounds that the Scottish Government would provide core funding, with which 10–15 jobs would be secured, and that national park status held the promise of achieving social and economic sustainability on the island (Harris National Park Study Group, 2010). The group requested the Scottish Ministers to "progress" their proposal for a national park as a "land-based park comprising the parish of Harris (excluding Berneray)",[34] with the structure of governance outlined as Option 2 in the feasibility study where the board would comprise 40 per cent of directly elected members, and with similar planning powers to those of the Cairngorms National Park Authority (Harris National Park Study Group, 2010) (Map 3.2). The group noted the positive meeting with Roseanna Cunningham, who had replaced Michael Russell as Minister for the Environment in August 2009, and her willingness to support the initiative in Parliament with an "indicative" annual budget of £1 000 000, providing that Comhairle nan Eilean Siar

Table 3.3 Land ownership on Harris.

		Area (ha)	Owner
Parish of Harris – mainland areas			
1	North Harris Estate incl. Soay Mor and Beg	24844	North Harris Trust
2	Amnuinnsuidhe Castle Estate	450	Ian Scarr Hall
3	West Harris	6604	West Harris Crofting Trust
4	Bays of Harris (incl. Stocknish Island)	10688	Rodney Hitchcock
5	Rodel Lands	810	Donald MacDonald
6	Kyles	405	Barry Lomas
7	Kyles Lodge	10 (est.)	Tom Jourdan
Parish of Harris – offshore islands			
8	Scalpay (linked by bridge)	703	John Taylor
9	Scarp	1045	Anderson Bakewell
10	Taransay, Gaisgeir and Gaisgeir Beg	1503	Jowhn Mackay
11	Berneray (part of Bays of Harris)[i]	1011	Rodney Hitchcock
12	Ensay (incl. surrounding islets)	186	Mackenzie
13	Killegray	176	Harry Wolfe
14	Pappay	820	David Plunkett
15	Shillay and Little Shillay	55	Andrew Johnson
16	Coppay	11	SGRIPD[ii]
17	Small islands in the Sound of Harris	39 (est.)	No owners identified
18	St Kilda (Boreray, Soay, Hirta)	855	National Trust for Scotland
19	Rockall (incl. Hasselwood Rock)	0.7	No owner identified
Total (CnES cites Harris as 50090 ha)		50250	

[i] Berneray is included for completeness as it is technically part of the parish of Harris. However, recent infrastructural works have strengthened the socio-economic links with North Uist. For this reason the study group does not consider that Berneray can reasonably be included within a National Park unless the people of Berneray decide otherwise.
[ii] Scottish Government Rural Payments and Inspections Directorate.

Source: Bryden *et al.*, 2008: 20. Used by permission of the Isle of Harris National Park Study Group.

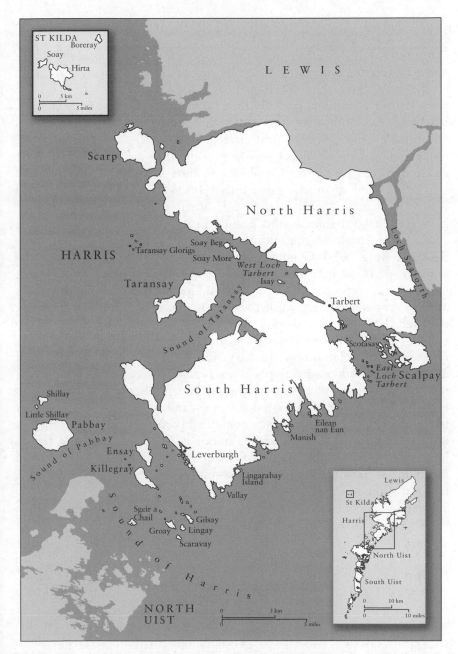

Map 3.2 Proposed Isle of Harris National Park. Based on original by HelenStirling.com. Used by permission of the Isle of Harris National Park Study Group.

gave its support before the deadline for the next spending review (Harris National Park Study Group, 2010).

In the face of overwhelming local support for the proposal on Harris and accumulated evidence, the Comhairle remained sceptical. In addition to the feasibility study (Bryden *et al.*, 2008), a presentation on the proposal for a national park by the Harris Study Group, with Duncan Bryden (lead author of the feasibility study), to the Comhairle's Sustainable Development Committee in April 2009, and a subsequent, much delayed, meeting between representatives of the Harris study group and members of the Comhairle in June 2010, the Comhairle had sought the view of others outwith the Outer Hebrides. These included Roseanna Cunningham, Highland Council members of Cairngorms National Park, Highlands and Islands Enterprise and local authorities connected to both existing national parks.[35] As discussed in the report dated 21 April 2010 (Comhairle nan Eilean Siar [CnES], 2010a) of the Director of Development of the Comhairle's Sustainable Development Committee, responses to a questionnaire distributed to the local authorities revealed that, while they were conscious that it was too early to reach definitive conclusions, they were "generally supportive of the benefits of national park status". The local authorities noted the generally good working relationships between the local authority and the park authority in both national parks and positive economic benefits as measured, for example, by an increase in the number of jobs created, an expansion in tourism and related activities, and an increase in investment in the area (CnES, 2010a). Two councils noted that the availability of affordable housing had become "a major issue" as a result of a rise in house prices, and Argyll and Bute Council stated that proposals for wind farms and aquaculture had been "discouraged" (CnES, 2010a) – although no indication of scale of these proposals is mentioned.

Despite the generally favourable responses from those approached, the Director of Development, Sustainable Development Committee, in a report dated 27 October 2010, recommended that the Comhairle "inform" the Scottish Government that, "at this stage a convincing case [had] not been identified for the creation of a National Park in Harris" (CnES, 2010b: 1). The "case", according to the report, had very clearly to be demonstrated on the grounds of the economic contribution that would result from national park status, and the case that had been made, the report asserted, was far from convincing: "In summary, while there is relatively positive anecdotal comment in regard to existing National Parks and their economic impacts, there is (at this stage) little hard, statistical evidence to demonstrate economic gain" (CnES, 2010b: 8).

"The special properties of the Harris landscape and environment", the report notes, "are already recognized by national and international designations" (CnES, 2010b: 7), and there is "concern" about "over designation" (CnES, 2010b: 8). The door was, however, not completely closed. The report also recommended that Scottish Ministers

> proceed to Stage 2 if their view is that a case exists to establish a National Park in Harris and in line with the legislative process, the Comhairle would provide a final definitive view at Stage 3, when the critical components of the National Park were known (CnES, 2010b: 1).[36]

In the Comhairle's Director of Development's report, the national park designation is constructed exclusively as an "environmental designation", whose meaning is bound discursively to such designations as the Special Site of Scientific Interest, the Special Area of Conservation and the Special Protection Area in North Harris (CnES, 2010b: 7) – despite the citation of the four aims of the legislation earlier in the Committee's report. "Local experience", the report continues, "suggests that designations will be utilised to either restrict or prevent legitimate development" (CnES, 2010b: 7). National parks, the report concludes, "are primarily concerned with natural heritage" (CnES, 2010b: 8).

As in the case of debates in the Comhairle over the proposal to give planning permission for the superquarry at Lingerbay in South Harris in the mid-1990s (Barton, 1996; Owens and Cowell, 1996; Mackenzie, 1998b; McIntosh, 2004),[37] the report and the majority of councillors debating the proposed national park cast the debate in terms of jobs versus environment, "jobs blackmail" to Laurie Adkin (1992; 146).[38,39] This was yet another environmental designation where ecological "preservation" effectively inoculated the land against economic "development", a case of "conservation aspic", to cite the pithy words of a *West Highland Free Press* Editorial (*WHFP*, 28 January 2011: 15). Threatening a formal "no" campaign on Harris itself, Donnie Macdonald, owner of the land where the Lingerbay superquarry was to have been located, similarly drew on this construction to legitimate his position (cited in *WHFP*, 27 February 2009: 3). He was "convinced" that more environmental designations would follow national park status with a resulting impact on economic activity, including any future development of the Lingerbay quarry site and renewable energy projects (*WHFP*, 27 February 2009: 3). Subscribing to a norm of nature as separate from the human – on which binary the jobs versus environment debate depends – opponents were blinded to the possibilities that could emerge from disturbing this norm, possibilities latent in the National Parks Act and evident in the practices of the two existing national parks. What Franklin Ginn (2008: 350) has referred to in the context of Aotearoa New Zealand as the "preservationist paradigm", which he considers "has done so much to solidify the Western separation of nature and culture", is clearly playing itself out here against a different set of political and ecological possibilities.

Conclusion

My intention in this chapter has been to demonstrate how community land ownership has the potential to undo a norming of nature as separate from the social, challenging the ways in which nature is given to be seen by either conservationist interests or those of private land owners. Relating local action to national and global narratives, I show how a community land owning trust's planting of native woodland, a management agreement with SNH and a proposal for a national park work together to reconstitute a nature – and a wild – whose meanings interrupt "the colonial legacy" (Toogood, 2003: 167) of statutory conservation and counter the construction of "emptiness" of the class of private land owners. All three initiatives insist that nature does not lie outwith history – and politics – but is bound up with their creation. They provide a key to re-imagining a politics of nature of which humans are part. To follow Butler's (2004) line of argument – developed by disturbing the category "human" and the notion of the body – this new politics has to do with keeping open to continuous resignification the meanings of "nature" and "wild". Nature, as the body in her analysis, is not "a static and accomplished fact, but … a mode of becoming that, in becoming otherwise, exceeds the norm, reworks the norm, and makes us see how realities to which we thought we were confined are not written in stone (Butler 2004: 29). Nature is still a-working. This new politics is, then, "a politics of contingency" (Moore et al., 2003: 4), co-produced through the inter-working of non-human and human nature.

The planting of native woodland gestures towards this new politics in several ways. First, the planting disrupts the "firm categories" (Cruikshank, 2005: 167) of nature and the wild by exposing the multiple and interwoven narratives through which the land has been and continues to be produced. There is, as I have shown, no original landscape to which nature or the wild may be secured, no pristine "wilderness" that exists outwith human history. Visually, the action of a spade, digging a hole for a seedling and thereby revealing the roots of a time-distant woodland, troubles the production of the present tree-less landscape as "wild", as the norm. It interrupts the class-based claim to a wild that serves the interests of private land owners – the preservation of the land for blood sports and the estate as a globally traded commodity. It calls into question any historically stationary constructions of a nature that might authorize conservationists' actions.

Second, the seeds of the native trees – however precisely local their provenance – are enrolled in this counternarrative. Confined until recently to the precipitous edges of the land where they could escape the foraging of sheep and deer, the seeds mobilize a biodiverse past to which they are genetically connected, a past where, the evidence suggests, people saw themselves – poetically and politically – as bound to a nature of which they were part. Discursively, they conjure a time when collectively held use rights to woodland were prevalent, governed through the law tracts of the tuath (Hunter, 1995a: 58–62). Planted as seedlings through the auspices of a community land owning trust, the seeds now re-engage with a collective ethic. They are integral to the contemporary exercise of collective rights to land and thereby to the creation of a more socially just politics (see Fitzsimmons, 2004: 44). At the same time, the seeds connect this locally informed collective ethic to globally sanctioned discourses of biodiversity, of action to mitigate the effects of climate change, and of sustainability.

Third, by opening the meanings of nature and the wild to new configurations, people, individually and collectively, reposition themselves with respect to nature. As Casid (2005: xviii) has argued with respect to the construction of counter-colonial landscapes in the Caribbean, the seeds and the woodland are implicated in "the growth of new subjectivities and a different sense of the political". Through the planting, the community trust, as a collectivity, positions itself as responsible for a nature of which it is part. There may be an attempt to be as "authentic" to the immediately local as the procurement of the seeds of native trees permits, but the evidence of interviews with trust employees, of participation in such events as guided walks, and in written reports, suggests that there is no attempt to create a nature or a wild divorced from historical or contemporary human action. The "natural heritage" and the "cultural heritage" are presented as intertwined. Both are implicated in this recent, collective, "doing" of the land.

The NHT's management agreement with SNH furthers this politics. The land in question is subject to multiple protective conservation desig-nations, a matter that remains unpopular among some in the crofting constituency (see MacDonald, 1998: 239). But now, subject in each case to limitations prescribed by legislation, those who live on the land have collective responsibility for its well-being. This includes responsibility for globally rare species and threatened habitats. As the trust has taken on this responsibility and reworked its positionality with respect to nature, so social boundaries have been reconfigured. The clearest example concerns deer management. In the case of the agreement between the NHT and the owner of Amhuinnsuidhe Castle Estate, deer stalking may remain a class-defined prerogative, but income from the lease of these rights accrues to the NHT and is thus tied to a collective ethic. More radically, the rights and responsibilities that the NHT has now delegated to the Stalking Club trouble further the boundaries that used to be exercised through class privilege. As I have explained, nature remains a commodity, but now that deer are once more held in common – and the fees minimal – stalking is no longer the exclusive domain of the wealthy. This counternarrative of stalking is supported through the club's commitment – aided by the NHT – to training in both best practice and deer biology. As in the case of the planting of native woodland, the management agreement with SNH is leading to local employment and the growth of local knowledge and expertise. The numbers may be small but they are significant in terms of the creation of skilled employment and the reskilling of the local population.

The NHT's leadership in proposing a national park whose perimeters extend well beyond its own estate provides evidence of a further gesture towards a new politics of nature. First, the initiative troubles a norm concerning the authority to instigate such a designation, responsibility in the past having rested with an outside statutory conservation agency. It is SNH, as Fraser MacDonald (1998) has shown, that has exercised the right in the past to define "natural heritage", *contra* crofters' claim to responsible stewardship of the land. A locally constituted land owning body is now the author of a proposal to establish a national park, a designation that would not displace such designations as Special Sites of Scientific Interest or Special Area of Conservation, but would potentially place the NHT in a position where it could be part and parcel of the negotiation of their meanings. The initiative, in other words, undercuts a colonial conservation optic (Mather, 1993; Toogood, 2003).

At the same time, and second, the NHT's proposal troubles the norm that so frequently underpins such designations – the separation of what counts as "nature" and what counts as "culture". The legislation pertaining to national parks is crucial here. While the objectives stated in the National Parks (Scotland) Act 2000 – supporting social and economic sustainability at the same time as sustainability of the "natural" and "cultural" heritage – do not preclude the emergence of conflicts that rely on the binary nature/culture, the legislation does allow for the more contingent working of these categories. In this respect, the Scottish legislation is aligned with international (IUCN) discourses that recognize the co-production of the social and the natural in the classification of national parks (Stockdale and Barker, 2009). With reference to the proposal for a national park on Harris, the feasibility study commissioned by the Isle of Harris National Park Study Group makes clear that this is a working land and points towards an opening of nature and the wild to non-essentialist configurations (Bryden *et al.*, 2008).

Third, together with disturbing the fixity of the binary nature/culture, people's repositioning of themselves *vis à vis* the land in a new politics of nature concerns practices of governance in the national park. As I have shown, these have to do with the powers granted through the legislation particular to each national park and with membership in the governing body, the national park authority. The authors of the feasibility study, Bryden *et al.* (2008), recommended "call in" powers and substantial, but not majority, local input, the latter on the grounds that it would support the local democratization of the process while recognizing the smaller population base than exists in the other two national parks.

Finally, in contrast to the planting of native woodland and the NHT's management agreement with SNH, discussion of the proposal for a national park on Harris has demonstrated that the process of resubjectivation may prove contentious. On the island, a clear majority expressed their support in a ballot in the face of a small but vociferous minority. In the Comhairle's debating chambers, this has not proved to be the case. At both levels, those opposing the proposal have positioned themselves "for" jobs and "against" the environment. Caught in the "givenness" of this construction, opponents have been unable to envision the possibilities that might emerge from disturbing the norm of the human/nature divide on which this construction depends. As a result, while not completely ruling out the proposal, the Comhairle prevaricated in its recommendation to the Scottish Ministers in 2010.

Each of the three moments of engagement with nature/the wild on which I have focused here provide the means through which property is performed with community land ownership. As I have shown in Chapter 2, a counterhegemonic narrative of property – of a commons, or land to which there is a common right – unsettles a neoliberal norm of privatization. It destabilizes an ownership model that relies on the dualism public/private. At the same time, opening up the land to new political possibilities through a reversal of property norms prompts a rethinking of the process through which nature has been normalized – through the separation of the non-human and the human – when land was held as a private estate. Community trusts' exercising of a collective responsibility towards the land, in other words, opens up nature to new positionalities and new responsibilities – of taking a position "*in* nature" (Braun, 2002: 13, emphasis his) and from this position asking questions about how nature might be reworked in the collective interest – for the commonweal – and with what consequences. As MacDonald (1998) and Hunter (1991) would insist, this position is one anchored in crofters' claim to time-honoured stewardship of the land. In the case of the NHT, the reworking of nature with community ownership – contested as this process sometimes is – suggests a more socially just becoming of nature, and of place, than had hitherto been the case.

Notes

1   McHardy (2010: 15) notes "a problem" with his line of argument, pointing out that "the word '*Cailleach*' is only thought [on linguistic grounds] to date from the Iron Age", not the Neolithic. Nevertheless, he considers that the discrepancy might be explained by the fact that many Neolithic sites, particularly those in Atlantic Scotland, were re-used during the Iron Age (McHardy, 2010: 15).

2   The archaeological research by Gregory *et al.* (2005: 944) in Northton, South Harris, is important as it provides direct evidence for the first time of Mesolithic occupation in the Outer Hebrides and of these islands' position as part and parcel of the "European Mesolithic mainstream" (Gregory *et al.*, 2005: 944; also see Armit, 1996). Of additional importance for the argument I am tracing here is the researchers' evidence of how Mesolithic activity was implicated in the formation of the soil in such a way as to influence subsequent settlement patterns and sites of agricultural activity, in other words, as evidence of the porous boundary between the natural and the social.

The palaeoenvironmental research by Edwards *et al.* (2005) specifically focuses on soil formation on the machair, found on the Atlantic facing coasts of the Outer Hebrides. Comprising "the crushed fragments of marine mollusca and crustacea as well as quartz sand and other products derived from reworked glaciofluvial deposits" (Edwards *et al.*, 2005: 435), the machair is recognized as providing some of the most fertile soil for agricultural purposes. While the authors do not contest "the natural origins" of the machair – attributable to the work of wind and waves following a rise in sea level during the Holocene and the location of large deposits of sand in the shallow waters of the adjacent Atlantic (Edwards *et al.*, 2005: 446) – they suggest that anthropogenic activity played a part in the machair's development. In both pieces of research, the soil, so frequently cast as a given physical entity, is shown to be bound up with social process.

3   The recent controversy over the proposal to site such pylons on a 137 mile route between Beauly and Denny, intruding – as many see it – on a much valued area of wild land in the Highlands, is but one particularly politically prominent political instance (see, for example, McDade, 2010).

4   The information cited here came from a series of interviews with Murdo Morrison (22 November 2000, 29 June 2001, and 7 August 2002. Also see Cunningham, 2000: 5; Fletcher, 2002: 10). Funding for the project came from Millennium Forests Scotland, Forest Enterprise, Western Isles Enterprise and SNH (interview, Murdo Morrison, 22 November 2000).

5   The NHT owns the land, except for 5 m on either side of the main stem of the Langadale River, which was sold by Ian Scarr Hall to the owner of the Grimersta Estate. Salmon spawn in this part of the river. Fishing takes place in Loch Langabhat, to which four estates – Morsgell, Grimersta, Ath Linne, and Scaliscro – have rights (NHT Land Manager, 31 March 2011).

6   The survey (SNH, 2008b) identified downy birch, aspen, hazel, holly, sallow willow and rowan at the Rhenigeadal and Loch Seaforth site. Such is the richness in biodiversity of this site that SNH (2008b: 29) has listed it as one of their four Core Areas of interest for native woodland. Two other sites are identified in the 2008 survey, both in South Harris, as sources of seed (or, in the case of aspen, of cuttings): an area of coast between the Tairbeart jetty and Direcleit and the islands and banks of Loch Plocrapoil. A number of sites in South Lochs – Loch Seaforth, Isle of Lewis, are named as Core Area 3 in terms of interest for native woodland and could provide further sources of seed for the plantings in Harris (SNH, 2008b: 29).

7   In 2000, it was estimated that native woodland covered about 150 000 ha of land, comprising 2 per cent of Scotland's land area. Of this, 88 per cent was found in the Highlands (Forestry Commission Scotland, 2000: 25–26). *The Scottish Forestry Strategy* (Forestry Commission Scotland, 2006: 16) notes that native tree species occupy 29 per cent of the area under forest; of this, 45 per cent are defined as "semi-natural".

8   Comhairle nan Eilean Siar's *Ar Nàdar. Frèam airson Bith-ionadachd sna h-Eileanan Siar/Our Nature – A Framework for Biodiversity Action in the Western Isles* (2004) was followed in 2005 by a series of specific Local Bio-diversity Action Plans (LBAP). For example, see Comhairle nan Eilean Siar, 2005d. Native woodland is also the subject of a specific Local Habitat Action Plan (LHAP).

9   This policy document replaced the Forestry Commission Scotland's *Forests for Scotland* (2000). In contrast to the earlier document, *The Scottish Forestry Strategy* (2006) paid far greater attention to climate change, to a more integrated approach to forestry – including native trees and biodiversity – and "community development". *The Scottish Forestry Strategy* (2006: 64) defines native woodland as "woodland that wholly or mainly

comprises species that colonised Scotland after the last Ice Age and before human influence on natural processes became significant".

10   The planting of native woodland is also supported by funding available, for example, from the Forestry Commission's Scottish Forestry Grants Scheme, a Highlands Locational Premium, a Western Isles Locational Premium, the Millennium Forest for Scotland Trust and SNH. The Scottish Forestry Grants Scheme provides 90 per cent of the cost of each scheme that focuses on the expansion of native woodland, the restoration of land where native species grow and improving riparian habitats (Forestry Commission Scotland, 2005: 3–6). In the Outer Hebrides, "top-up" funding was provided by the Western Isles Locational Premium between 2004 and 2008, provided by the Scottish Executive (Forestry Commission Scotland, 2004).

11   Paralleling the argument to deconstruct the binary nature/culture, Warren (2007) argues that the terms "native" and "alien" need to be unpacked and their politics unmasked. He views the construct native/alien as "spatiotem-porally arbitrary", as having "disturbingly xenophobic associations", and as being logically and ethically problematic (Warren, 2007: 427). In their place, he calls for "a more discerning, pragmatic approach based not on arbitrary and contested ideas of naturalness but on the socio-biogeographical specifics of place, context and species" (Warren, 2007: 437). He suggests that "a species' potential for causing harm in a particular place and time may be the most useful, honest and ethically defensible criterion for guiding conservation choices" (Warren, 2007: 437). This he refers to as a "damage criterion" (Warren, 2007: 427).

12   For a summary discussion of the origins of A'Chraobh Ogam, see Mackenzie (2002).

13   For SNH, "wildness" refers to the "quality enjoyed", whereas "wild land" denoted "places where wildness is best expressed" (SNH, 2002: 1). "[W]ild land has normally been identified in the remoter areas in the north and west" of Scotland, continues SNH (2002: 1), whereas "the quality of wildness can be found more widely in the countryside, sometimes quite close to settlements".

14   While SNH states that "[w]ildness cannot be captured or measured" (2002: 1), the organization lists five physical attributes that "contribute to the experience of wildness" and that can be "recorded and assessed by a simple scoring":

    –   a high degree of perceived naturalness in the setting, especially in its vegeta-tion cover and wildlife, and in the natural processes affecting the land;
    –   the lack of any modern artefacts or structures;
    –   little evidence of contemporary human uses of the land;
    –   landform which is rugged, or otherwise physically challenging; and
    –   remoteness and/or inaccessibility (SNH, 2002: 13).

It is the last attribute that is then used to map wild land, first using distance from a public road (at 2, 5 and 8 km) and, second, distance from "any motorable road" (SNH, 2002: 7). "Evoked" by these physical attributes, SNH notes four "perceptual responses":

- – a sense of sanctuary or solitude;
- – risk or, for some visitors, a sense of awe or anxiety, depending on the individual's emotional response to the setting;
- – perceptions that the landscape has arresting or inspiring qualities; and
- – fulfilment from the physical challenge required to penetrate these places (SNH, 2002: 13–14).

15 It is important to note that the National Scenic Area designation is not confined to wild land. It has a broad definition, which encompasses "[n]ationally important areas of outstanding natural beauty" (Warren, 2009: 221). National Scenic Areas cover 12.7 per cent of Scotland (Warren, 2009: 221). SNH's (2008a) *Guidelines for Identifying the Special Qualities of Scotland's National Scenic Areas* now provides a methodology for professionals to use in identifying the "special qualities" of landscape and scenery that led to an area's designation. The rationale is that this identification will lead to more robust management strategies (SNH, 2008a: 2).

16 As I am drawing on some sources with restricted distribution, I do not identify the rivers by name.

17 A total ban on pearl fishing has been in force since 1998. Since that time, pearl fishing has declined by 90 per cent in Scotland (Hastie, 2004: 10). However, writes Hastie (2004: 10), there is a group of people who continue to fish illegally, and isolated mussel beds such as those in North Harris are particularly vulnerable in this respect. "Kill sites" were recorded as recently as 2009 in Harris, Skye and Lochaber (*Stornoway Gazette*, 9 April 2009: 3). Also see Cosgrove *et al.*, 2000.

18 Red deer are the only species of deer in the Outer Hebrides and the introduction of other species is prohibited. Unlike red deer on the mainland, there has been no interbreeding with the introduced Sika (see Warren, 2009: 299–305). The NHT is currently participating in genetic research, which may indicate the provenance of the herd. The most recent count, in 2010, indicated a reduction in numbers from 2008 when 1377 deer – 359 stags and 1018 hinds/calves – were counted. The more recent count indicates, according to the Land Manager, that the numbers are now "under control" (North Harris Trust Land Manager's report, 2010: 24).

19 With the buyout of the Loch Seaforth Estate, the NHT adjusted the boundaries. The Ath Linne Estate was allocated land on which to stalk in Maaraig and Scaladale, and Caolas was removed. Rights to Caolas were given to the club, together with those of Reinigeadal.

20 The NHT Land Manager provided the information regarding deer management (interview, 15 November 2010).

21 The aim is to reduce deer numbers in order to allow vegetation to regenerate, which in turn will provide cover for grouse and hare and lead to a more sustainable supply of food for eagles (interview, NHT Land Manager, 28 March 2011). For a fascinating study of the relationship between eagle diet and successful breeding in North Harris and the Eisgen Estate on Lewis, see Reid (2008).

22 The figures for 2010 for Amhuinnsuidhe Castle Estate were 10 hinds and 30 stags.

23 The NHT contributed to the cost of training through its Community Development Fund. Additional funds have come from working on the deer count (£75 per person per day) and each trainee paid £50. The club hired its own trainer (interview, NHT Land Manager, 15 November 2010).

24 Like SNH, a national park authority is a "Non Departmental Public Body". As such, it has responsibility for "administrative, commercial, executive [and] regulatory functions on behalf of the Government" (Spencer, 2008: 7). It has its own budget and receives "grant-in aid" from the government for its operation (Spencer, 2008: 7). Its employees are not classed as civil servants (Spencer, 2008: 7).

25 Particularly in the case of the Cairngorms National Park, initial hostility on the part of some local authorities was overcome by the insistence that this designation "should be seen as much as an economic tool for the area as protecting its natural heritage" (interview, John Randall, 3 July 2008). It was interesting, commented Randall, that "there was a big shift in opinion [on the part of local authorities] in the Cairngorms from hostility to seeing this as an economic opportunity" (interview, 3, July 2008). Once this was realized, he continued, "people [were] falling over themselves to get in" (interview, John Randall, 3 July 2008).

26 This stipulation, that greater emphasis should be placed on the first aim where there is conflict, parallels what is referred to as the Sandford Principle, an outcome of the National Parks Policy Review Committee 1971–1974 for England and Wales (Stockdale and Barker, 2009: 481).

27 The Countryside Commission for Scotland (1990) proposed four national parks: Loch Lomond and the Trossachs, the Cairngorms, Glen Coe – Ben Nevis – Black Mount, and Wester Ross.

28 A two-stage model of enactment of national parks was followed. As the primary legislation, the Act of 2000 establishes the basic framework for national parks. Secondary legislation provides for the establishment of each park through designation orders specific to it (National Parks [Scotland] Act 2000, Sections 6, 7). In practice, this means that the Act "allows for a significant degree of flexibility and interpretation" (Stockdale and Barker, 2009: 482) and that, thus, the secondary legislation can be "fine tuned" to meet the specific objectives of each park (Warren, 2009:

237). The legislation followed an extensive period of consultation (Warren, 2009: 239).

29 In March 2008, the Minister responsible, Michael Russell, made the decision not only to adjust the park boundary such that it included Highland Perthshire but also to review the park authority's operating powers (Stockdale and Barker, 2009: 490).

30 The study identifies a number of options regarding park boundaries, including one for the whole Parish of Harris, of which, historically, St. Kilda has been part. Bryden *et al.* (2008: 22) remark that, although the inclusion of St. Kilda "would add profile to a Harris National Park" owing to its "iconic status", it was already "managed in the public interest" and it "could bring considerable practical challenges and possible additional costs". Their preference would be for the proposed park to include mainland Harris and the three "most closely connected" and largest offshore islands – Scarp, Scalpay and Taransay (Bryden *et al.*, 2008: 21).

31 Bryden *et al.* (2008: 18) also include St. Kilda in their discussion at this point.

32 Following a pattern established by the Cairngorms National Park, four of the 15 staff positions would be concerned with corporate services, three with heritage and land management, two with visitor services and recreation, three with economic and social development, two with planning and one with strategy and communication (Bryden *et al.*, 2008: 54). The study estimates that an annual core budget of £800000 would include £500000 for the 15 staff positions, £200000 for projects and £100000 for property rental and overheads (Bryden *et al.*, 2008: 54).

33 On the basis of their research, the Spencer review comments that it is widely acknowledged that the board's effectiveness with respect to local interests is enhanced through the inclusion of directly elected members and that there is "some support" for their increased proportional representation (2008: 7), recognizing at the same time that "national parks were not "local" parks" (Spencer, 2008: 29).

34 In contradistinction to the writers of the feasibility study, the study group recommended the park comprise the entire parish of Harris (i.e. including St. Kilda), excluding Bernera, which is now linked by causeway to North Uist and is more tied to that island socially and economically (Harris National Park Study Group, 2010). The people of Bernera, the group suggested, should make their own decision on the matter. St. Kilda was first named a UNESCO World Heritage Site in 1987 on the grounds of its "natural environment" (Ross, 2009: 12). In 2005, it was again named a World Heritage Site, this time on the grounds of its "unique cultural heritage". It is the only dual World Heritage Site in the UK, one of a global total of 25 (Ross, 2009: 12). Commenting on the inclusion of St. Kilda, Duncan MacPherson, NHT Land Manager, stated: "At the public meetings people very much felt St Kilda should be part of the national park. St Kilda

is part of the parish of Harris. Its history belongs to the folk of Harris"
(cited in *Press* and *Journal*, 23 April 2009: 6).

35 In addition to evidence provided from these sources, the Comhairle's
Director of Development draws on a report from the Association of
National Parks annual conference attended by the Vice-Chair of the
Sustainable Development Committee and a Comhairle employee. It is
interesting – and disturbing – to note that, in the reported feedback from
the conference, no distinction was made between national parks in
Scotland on the one hand, and those in England and Wales on the other
(CnES, 2010b).

36 Following the requirements of the legislation, Stage 1 refers to the submission of a proposal for a National Park, Stage 2 involves "consideration and
production of a statement" and Stage 3 concerns consultation (CnES,
2010b: 8).

37 The proposal for a superquarry at Lingerbay was finally dropped by
Lafarge Aggregates in April 2004, three months after the Court of Session
ruled against that company in the matter of the validity of a 1965 licence
and ten years after the start of the public inquiry, which lasted over eight
months (*WHFP*, 9 April 2004: 13).

38 I draw here on discussions with the NHT Land Manager, 20 January 2010
and 3 August 2010.

39 One interviewee considered that some Comhairle councillors would have
opposed the proposed Harris National Park because of the inclusion of
St Kilda. While Harris included St Kilda on the grounds that it was part of
the parish of Harris, others may have seen its inclusion as a threat to the
outcome of an earlier struggle (2009) between three communities – North
Uist, Harris and Uig (Lewis) – to site a St Kilda Centre. The initiative came
from Comhairle nan Eilean Siar, Highlands and Islands Enterprise, Visit
Scotland, the Gaelic Arts project and the National Trust for Scotland
(*WHFP*, 6 November 2009: 15). Amidst substantial controversy, on the
consultants' recommendation, the award went to Uig, the one site of the
three with no historical attachment to St Kilda.

4

Working the Wind

Complicating commodification

The Dancing Ladies of Gigha, as the three wind turbines on that island off the Mull of Kintyre are affectionately known, have become an icon of success for other communities aspiring to land ownership. Recognizing that a secure source of income was essential for turning around decades of population loss and economic decline, the Isle of Gigha Heritage Trust, which took over ownership of the island in March 2002, placed its hopes on generating electricity from the wind and selling it to the national grid. All income, once loan repayments and running costs are paid, is invested in the community, new housing and the upgrading of existing housing being among the early priorities. Speaking on the occasion of the launching of the wind farm in January 2005, Willie McSporran, Chair of the Trust and Director of the Gigha Renewable Energy Company, emphasized the significance of the wind farm for moving towards a more sustainable future. "Until now", he said,

> nobody had managed to crack the financial nut enabling a community with little money to become a significant local generator. The solution we have developed in Gigha works by combining grant funding with loan and equity finance secured at commercial rates. The company simply pays back the loan and buys back the equity within five years. What's more, by year eight we will have built up a capital reinvestment fund sufficient to replace the machines without recourse to further financing (cited in *WHFP*, 28 January 2005: 2).[1]

Places of Possibility: Property, Nature and Community Land Ownership,
First Edition. A. Fiona D. Mackenzie.
© 2013 A. Fiona D. Mackenzie. Published 2013 by Blackwell Publishing Ltd.

Other communities have followed Gigha's lead, Eigg being the most widely quoted example. This island achieved energy self-sufficiency in 2008 on the basis of its grid system run by Eigg Electric, a wholly owned subsidiary of the Isle of Eigg Heritage Trust which had bought the island in 1997 (*WHFP*, 8 February 2008: 1). The grid is powered by a combination of wind, micro- and larger-scale hydro schemes and photovoltaic panels. Previously, with no mains electricity available, most households had relied on individual diesel generators to produce electricity, a source both expensive and environmentally problematic (*WHFP*, 1 June 2007: 2). Income from the scheme is used to employ five local people, whose task is to maintain the system (*WHFP*, 9 January 2008: 1). Eigg received Scottish Renewables' Green Energy Award in 2009 in recognition of its success in achieving energy self-sufficiency through renewable energy. In 2010, it won a £300 000 prize from the *Big Green Challenge*, a competition organized by the UK National Endowment for Science, Technology and the Arts, for its work on a series of projects that included house insulation, fitting photovoltaic panels on houses and public buildings and expanding the use of bicycles (Community Energy Scotland, 2010: 5). Globally, Eigg has become an icon of what can be achieved locally with respect to achieving energy self-sufficiency through renewable sources and for cutting carbon emissions by almost 50 per cent.

Encouraged by these stories of success and the support of the widely popular Community Energy Scotland,[2] a number of community trusts in the Outer Hebrides have sought planning permission for small-scale wind farms. However, the process of building a community economy on the basis of renewable energy has proved to be far from straightforward. Touted as one of the three groups of "energy islands" (the other two being Shetland and Orkney) with "world class energy regimes"[3] that are viewed as pivotal in terms of meeting UK and Scottish renewable energy targets,[4] the Outer Hebrides have become the target for corporate and private prospecting. Supported by the Comhairle nan Eilean Siar's economic policy that promotes the islands as an Energy Innovation Zone – a policy congruent with the government's "private developer/ public subsidy model" supporting renewable energy (Warren and McFadyen, 2010: 206)[5] – the search is for sites where electricity may be generated from the wind and, to a lesser extent as yet, from waves and tides. As examples which I discuss below, Lewis Wind Power, an AMEC plc and British Energy plc venture, proposed a 234 turbine wind farm, and Beinn Mhor Power, owned by a private individual, Nicholas Oppenheim, sought planning permission for a wind farm of 133 turbines in 2005, both initiatives on the Isle of Lewis.[6]

As I show, the process of commodifying the wind – the enclosure of a commons through corporate or private individual interest – proceeds according to the norms of the commodification of nature. The main beneficiaries are either shareholders in a registered company or a private individual's bank account. The planning process is crucial in the process of normalization. Through the requirement to provide detailed specifications of the technological components of the wind farm (turbines and associated infrastructure) and the satisfaction of criteria that include environmental, economic and social impact assessments on the one hand and the construction of debate about planning consent in terms of jobs versus the environment on the other, the planning process depoliticizes the process of enclosing the wind. Silently, it removes the deeply political question of land ownership from the discursive register.

The planning process operates as a political technology (Foucault 1979), separating rights to the wind from rights to the land. It renders the wind politically inert in the search for means to satisfy renewable energy targets, to cut carbon emissions and to achieve more sustainable futures. Through these signifiers – in a case that parallels the proposal to construct a super-quarry at Lingerbay in South Harris a decade earlier – what former Member of Parliament for the Western Isles Calum MacDonald calls "the 'big bang' approach" to "development" (cited in *Stornoway Gazette*, 17 July 2005: 3) proceeds. The establishment of a community fund or community ownership of a small proportion of the wind turbines as a condition for planning consent, to follow this line of argument, rather than mitigating the effects of privatizing the wind, becomes the means through which privatization proceeds.

The immediate effect of the granting of planning permission is to raise the value of the land, pushing it beyond the financial reach of community purchase. This process, together with the (common) practice of private owners to separate "development rights" to the land from rights to the land itself – through an interposed lease, which I discuss later – works against the intent of the Land Reform Act. In a situation where a community aspires to land ownership, the capture of the wind through corporate or private means, legitimated through the planning process, may thus clearly be named as "accumulation by dispossession" (Harvey 2003). "[L]andscapes of power", to use the language of M. J. Pasqualetti, P. Gipe and R.W. Richter (2002), in other words, further the dispossession of local people's rights to the land. "Welcome to the 21st Century Clearances", wrote one of those opposing Lewis Wind Power's proposal in a letter to the Comhairle (cited in CnES, 2005a: 315).

The process of dispossession is disrupted, as I show through the examples of the Galson Estate Trust and the Pàirc Trust, also on the Isle of Lewis, through the struggle to bring the land in question into community ownership. I examine how these two communities trouble the norms of the commodification of the wind as property rights are re-imagined in complex ways. In the subsequent section of the chapter, the argument is extended by exploring how a community land owning trust produces nature in its struggle to establish a small-scale wind farm in the face of opposition of SNH, the statutory conservation authority. I show how, through this new, collective, performance of property, the NHT reworks the idea of nature as external to social process – through which binary on this occasion SNH regulated the protective environmental designations that "enclose" North Harris and that could effectively dispossess a community of its claim to the wind. I argue that it is this articulation of nature with social process (specifically here community land ownership) that allows the community body to create the possibility of a different politics of nature, which points in the direction of more socially just engagement with "the new energy commons" (the phrase is that of Dan van der Horst and Saskia Vermeylen, 2008). In the final part of the chapter, I trace how, through this politics of a more social nature conjured through the wind, inchoate though this politics is in the majority of cases, people renegotiate their collective subject positions with respect to the economy.

Debating the wind: Comhairle nan Eilean Siar, Stornoway, 28 and 29 June 2005

The planning officer for the Outer Hebrides, Alastair Banks, brought forward for consideration, first by the Environmental Services Committee on 28 June 2005, and then by the whole Council on 29 June 2005, two proposals for large-scale wind farms. These were the first to be discussed by the Comhairle and marked the culmination of a process that had excited polarized debate in the islands as well as further afield. The first proposal was that of Lewis Wind Power Ltd (LWP), for a wind farm of 234 turbines to be located on the Barvas (63 turbines), Galson (97 turbines), and Stornoway Trust (74 turbines) estates. The wind farm would stretch from the west of Stornoway across Barvas Moor to Barvas, thence roughly north east to Ness, skirting the Moor on the east and numerous settlements to the west. It would have an installed capacity of 702 MW, making it the largest onshore wind farm in Europe; each 3 MW turbine would reach a blade tip height of 140 m; 167 km of access roads would be required, together with 32.5 km of overhead lines on 141 pylon towers, each 27 m high (CnES, 2005a: 3). The second proposal was for a 133 turbine wind farm by Beinn Mhor Power Ltd (BMP) in the south of Lewis, on the Eisgen Estate. The Muaitheabhal Wind Farm, as it is named, would have an installed capacity of 399 MW, each 3 MW turbine estimated to reach a tip height of 121 m; approximately 77 km of access roads were planned within the site, with the upgrading of other existing roads (CnES, 2005b: 3, 5).

In order to trace how the wind is produced as a commodity detached from the politics of the land, I draw on Banks's lengthy analysis of the two proposals presented to the Comhairle at the end of June 2005. The report on Lewis Wind Power's proposal runs to 627 pages, including appendices, that for Muaitheabhal to 376 pages. Key sections reflect on the two proposals' environmental statements – landscape and visual impact, including calculations of Zones of Visual Influence, ornithology, habitats, fauna, hydrology, geology and hydrogeology (including risk assessments for peat slides),[7] fisheries, archaeology and cultural heritage, safety, health and the environment, transportation, aviation, Ministry of Defence and telecommunications, socio-economics (including tourism and recreation) and renewable energy targets. In each case, there is detailed discussion of LWP's and BMP's environmental statements and responses from such consultees as SNH, the Scottish Environment Protection Agency and the Royal Society for the Protection of Birds (RSPB) in light of environmental designations, policy statements and planning guidance. Other submissions, including those from affected community councils, are provided as additional evidence. I also refer to the debates that took place among the councillors who were members of the Environmental Services Committee on 28 June and the full council on 29 June 2005. I attended the discussions on both occasions, one among many in a public gallery filled to capacity.

In his opening address to the Environmental Services Committee, Banks established the parameters for debate – the need to be cognisant in reaching a decision of both the "social fabric" of the islands (specifically, jobs) and the environmental issues raised by the proposed wind farms, both being of central significance to the development plans for the Western Isles, including the Structure Plan (CnES, 2003). He reminded the committee that it was their responsibility to consider the two projects "on their individual merits", not to discuss them with reference to alternatives, an admonition that, if followed, would remove from the discursive field any consideration of the possible effect of the granting of planning permission for these two large-scale wind farms on both wind farms led by communities that were then in the process of applying for planning permission and on those communities that had their sights on land ownership. He also brought to the attention of the committee members the matter of the sub-sea cable – the interconnector – which would be needed to take the electricity from the islands to the mainland. While this cable would be required whatever the scale of the wind farm, its justification on the grounds of economic efficiency could only be assured, he stated, if a minimum of 600 MW electricity was generated.[8] Banks recommended that, subject to specified conditions and the agreement of the Scottish Government, the Lewis Wind Power proposal be supported, but that of Beinn Mhor Power refused.

Banks distinguished between the two proposals on the grounds that, despite its "serious difficulties", LWP provided sufficient evidence of its potential to generate jobs to override environmental considerations whereas this was not the case with BMP's proposal. For reasons both of local economic exigency and national energy targets, Banks considered it vital to act immediately "to tap into a resource that is the best in the UK and possibly in the world". Using Department of Trade and Industry estimates for the consumption of electricity in 2010/2011, his report estimated that LWP's wind farm would produce

- 23% of Scotland's domestic power needs.
- 0.66% of the projected UK 2010 total electricity requirements.
- 7.09% of the projected total Scottish energy requirements.
- 6.38% of the UK 2010 Renewable Obligation Target of 10%; and 3.32% of the UK 2020 aspiration of 20%.
- 39.4% of the Scottish Target of 18%; and 17.73% of the Scottish 2020 target of 40% (CnES, 2005a: 372).[9]

There was a danger, Banks said, that the Western Isles could be left out as other sources of renewable energy, such as waves, were developed much closer to centres of consumption in England and Wales.

The key test in resolving "the serious difficulties" to which Banks referred – namely balancing the environmental cost with the promise of economic prosperity – was to be found in National Planning Policy Guidance Note 14 *Natural Heritage* (Scottish Executive, 1999a), paragraph 42, which refers to Natura 2000 sites. These are sites recognized by the Habitats Directive of the European Community and include Special Protection Areas (SPAs) and Special Areas of Conservation (SACs). Their objective, briefly, is to protect those habitats and species in member countries of the European Union that are deemed to be at risk. Paragraph 42 states:

A development which would have an adverse effect on the conservation interests for which a Natura 2000 area has been designated should only

be permitted where: there is no alternative solution; and there are imperative reasons of over-riding public interest, including those of a social or
economic nature (cited in CnES, 2005a: 81).[10]

Banks's hope was that the Scottish Government, to which a proposal of
this size had to be referred, would be persuaded by this argument. After
all, in addition to its promise of making a significant contribution to
national and UK targets for the production of electricity from renewable
sources of energy and its significance for the local economy, it would
drive the construction of an interconnector, which, in turn, would facilitate other renewable energy projects in the Outer Hebrides.

His case rested on the provision of jobs (Table 4.1). The Principal
Policy Officer (Economic Development) considered the figures for LWP
to be "reasonable" (CnES, 2005a: 308), but drew attention to a problem
with the predictions for the labour market. He noted LWP's conversion
of all categories of employment to full time equivalents (FTEs), despite
the fact that the construction phase would only last about four years and
that differently skilled workers would be needed during that period

Table 4.1 Summary of employment impacts at the Western Isles level – Lewis Wind Power (LWP) and Beinn Mhor Power (BMP). (Full time equivalents).

Source	LWP	BMP
Construction	133	80
Wind farm operation	46	25
Regeneration plan	–	23
Land rental payments	28	–[(i)]
Community fund and WIDT (over 25 years)	272	–
Eisgein Community Trust	–	9[(ii)]

Employment over 25 years from community ownership as a percentage of the wind farm

LWP	10%	15%	20%
	310	445	600
BMP	10%	17%	25%
	170	290	425

[i] As no land on the Eisgein Estate is under crofting tenure, there are no rental payments.
[ii] If the revenue to the Eisgein Community Trust is treated "as a general injection of expenditure into the local economy", 80 FTE could be created over a period of 25 years (CnES, 2005b: 347).
Source: Calculated from CnES, 2005a: 308, 309, 310; 2005b: 226, 346, 348.

(CnES, 2005a: 387). Many contracts would thus probably be of shorter duration than four years. He also noted that a number of areas remained open to further consultation, one such figure relating to the effects of a community fund (CnES, 2005a: 308). In their Environmental Statement, LWP states that it would contribute £560 000 annually to communities in the three estates and a further £702 000 to the Western Isles Development Trust (LWP, 2004: 8).[11] The Principal Policy Officer (Economic Development) identifies the far greater benefits to communities that would accrue from community ownership of up to 20 per cent of the wind farm (Table 4.1) (CnES, 2005a: 309).[12] Land rental payments are identified in Table 4.1 as a further source of community income. LWP would expect to pay an annual rent of about £4 million for the use of land belonging to the Stornoway, Galson and Barvas estates (CnES, 2005a: 283). Half of this would be paid to the shareholders of the common grazings on each estate, the remainder being divided among the owners as follows: 16 per cent to the Stornoway Trust, 21 per cent to the Galson Estate and 13 per cent to Barvas (CnES, 2005a: 283). As I show below, Galson Estate was at the time the subject of a hostile bid by the community to take over ownership of the land. Barvas has been considering community purchase of that estate. The Stornoway Trust has been in community ownership since 1923 and was the only trust of the three that has been unwaveringly supportive of the proposal (WHFP, 5 October 2007: 3).[13]

In contrast to LWP, Banks considered that the grounds for privileging jobs over the environment in the case of BMP for the Muaitheabhal Wind Farm were insufficient. Here, the key environmental test was found in Paragraph 25 of NPPG 14 Natural Heritage, which relates to the designation of National Scenic Areas. Forty of the 133 turbines of the proposed wind farm were to be located within the South Lewis, Harris and North Uist National Scenic Area. Paragraph 25 states that:

> Development which would affect a designated area of national impor-
> tance should only be permitted where (a) the objectives of the designation
> and the overall integrity of the area will not be compromised; or (b) any
> significant adverse effects on the qualities for which the area has been
> designated are clearly outweighed by social or economic benefits of
> national importance (cited in CnES, 2005b: 29).

Regarding environmental matters, Banks expressed concern about the potential for peat slides on the hilly terrain, referring to the disaster at Derrybrien, Ireland (see endnote 7). Further, the size of the wind farm would not, in itself, drive the interconnector, and the cumulative effect, if supported in conjunction with the Lewis Wind Farm and the proposal expected from Scottish and Southern Energy for a wind farm on the Pàirc Estate, might be too high and could compromise the chances of other proposals in the area. Here, contradicting his earlier caution to the committee, Banks stated that, while the future could not be allowed to "prejudice" the present, it was necessary to accept that the Pàirc Trust was awaiting a decision from the Scottish Government regarding its application to purchase that adjacent estate and bring it into community ownership.

SNH, as a required consultee, was persuaded by neither case. With respect to LWP, SNH stated in its communication to the Comhairle that – while it was mindful of its "balancing duties" as detailed in the Natural Heritage Act 1991, and thus would not object on the grounds of landscape – it was registering a formal objection on ornithological grounds and on the grounds of loss of habitat (cited in CnES, 2005a: 41). 190 (81 per cent) of the 234 turbines were to be sited within the Lewis Peatlands SPA, and it was anticipated that, of the 812 ha of habitat that would be lost through the development, 80 per cent (585 ha) would be from this designated area (CnES, 2005a: 59). Given that the SPA is roughly 58 984 ha, this would amount to a loss of under 1 per cent of the SPA, but this loss would potentially affect birds that are qualifying interest bird species for the SPA (CnES, 2005a: 59).[14] Coterminous with the Lewis Peatlands SPA is an area of blanket bog protected under the Ramsar Convention on Wetlands of International Importance Especially as Waterfowl Habitat. The Lewis Peatlands Ramsar Site qualifies under Criterion 1b as "one of the largest and most intact known areas of blanket bog in the world", under Criterion 2a by providing a habitat for rare wetland birds (including red-throated and black-throated divers, golden plover and greenshank) and under Criterion 3c for its support of a globally significant population of dunlin (CnES, 2005a: 126). Congruent with its assessment of the effect on the SPA, SNH concludes that it "cannot see how it would be possible for this development to proceed without the integrity of the Lewis Peatlands Ramsar Site being adversely affected" (cited in CnES, 2005a: 130). In the case of BMP, SNH expressed agreement with Banks's recommendation that planning permission be refused. Its formal objection noted the effects of the wind farm on landscape and "wild" land on the one hand and, on the other, the threat to golden eagles and white-tailed eagles (Letter to the Scottish Executive, 13 May 2005, included in CnES, 2005b: 248).

The majority of the councillors sitting on the Environmental Services Committee on 28 June 2005 and those present at the Council's debate the following day concurred with the planning officer's recommendation that the LWP proposal, subject to specific conditions, be sent to the Scottish Government with a recommendation that it be approved, despite overwhelming opposition from the communities affected. The majority of councillors disagreed with his negative assessment of BMP. Both projects, in their view, should be given planning permission. Congruent with the parameters set by Banks in his opening remarks on 28 June 2005 – that only the two proposals for large-scale wind farms be considered – supportive councillors subscribed to an "all or nothing" scenario suggested by this framing and a "jobs versus environment" casting of the debate. In both cases, claims of an "over-riding public interest" (LWP) or "outweigh[ing]" of environmental costs by "social or economic benefits of national importance" (BMP) in the achievement of a sustainable future for the Outer Hebrides centred on the proposals' promises of jobs. Some councillors also drew on nationally and globally anchored discourses of renewable energy and climate change to legitimate their positions.

To turn first to the debate on LWP, for one councillor, echoing the thoughts of others, this was a "fragile economy" and a society that was "slowly depopulating"; schools were closing; teachers were leaving. "Our islands may never recover", he continued, but the wind farm presented "an opportunity to bring new life into the islands". "We have to look at the big picture; we need to kick-start these islands; we need an injection of cash", said another. For another councillor, having lost the fishing and tweed, "this is all there is now; if we don't bring something in, there'll be nothing left". Another said he could not "see any future without the turbines; without them, we won't have these islands". A further councillor repeated the threat of a doomsday scenario: "without this [wind farm], we are facing a horrendous, horrendous, disaster". In response to a question about the possible negative effects on tourism, the major economic earner, and a net loss of jobs in the local economy as a result of the wind farm, another councillor suggested that, "it could be argued that additional tourists will come to see the turbines". "Will they come to see an empty landscape with no children playing?", he asked.[15] For others, it was the need for an interconnector that was the deciding factor. This, said one councillor, would not come with community wind farms. Addressing the issue of the high percentage of people against the proposal, another insisted that he was not "morally wrong" to take a position contrary to his constituents. His responsibility was one of "stewardship". Some people, he said, did "not understand the full implications of what's on the table". "Can anyone say South Harris is better off without Lingerbay?", he asked, a thinly veiled reference to the Comhairle's decision in 1993 to withdraw its support from that previous proposal for a massive development, a superquarry, that was touted by some as the solution to the islands' economic woes.

It was left to the eight councillors who voted against the proposal (19 supported it) on 29 June to call for "a balance between the economy and the environment", to cite the words of one, and to insert communities and the issue of land ownership into the debate. One councillor expressed amazement that, until that point in the discussion on 29 June, no one had mentioned communities. Yet in Ness, he said, an official ballot had indicated that 83 per cent of those eligible to vote opposed the wind farm, providing their representative with a very clear mandate.[16] His intervention met with loud applause. A second councillor who opposed the motion situated the proposal in the context of the politics of the land. When he first heard of the Lewis Wind Farm, he said, he saw benefits – for the islands and to counter global warming. The people with whom he spoke, he emphasized, were not against wind farms as such, but they were against the size of it. "We're sold down the river by landlords lining their pockets" – before communities have a chance to buy the land, he remarked. He continued., "We're no stranger to boom and bust – here we go again. ... People here are proud people who want to take control of their own destiny. ... Owning wind farms on our own is 100 per cent better than what we have on the table now. If we had a community wind farm, we could ensure the manufacture of turbines on the island and this would be important". To one of the councillors with whom I spoke after the debate, this was the new "Clearance" – corporately or individually led large-scale wind farms being the means through which community aspirations of land ownership, sanctioned through land reform legislation, could be pre-empted. This proposal and that of Beinn Mhor Power, she insisted, would promote "a dependency culture, when the wind provides the opportunity to do the opposite" (interview, 28 June 2005).

Councillors drew on similar supportive and oppositional discourses in the debate on BMP. For the majority, the promise of jobs was compelling and large-scale projects were the only solution. Said one councillor, "This is not about fiddling around at the edges. ... Putting in small schemes will not [deal with the issue]. It's not even relevant to bring small schemes into this argument". The councillor representing Lochs, the ward where the wind farm would be located, listed the "facts": population decline, from 4703 in 1901 to 1814 in 2001; "a beautiful environment", but this was "man-made", and "beauty alone does not put food on the table"; neither SNH nor the RSPB provided any employment in Lochs; the wind farm would provide "sustainable jobs" and a community owned wind farm; and, considerable development could take place without there being an impact on tourism. She wanted to see children playing there, she said, "not in photos on the mantelpiece'. The community was solidly behind the project, she insisted, citing figures of 118 opposed and 847 people either supporting the project or not feeling "strongly enough to object". Another councillor referred to "the designation blight" that resulted from being within the South Lochs, Harris, and North Uist National Scenic Area. "Where do we draw a limit because all of Harris is in a National Scenic Area?", he asked. The potential for a peat slide was mentioned by some councillors, but dismissed by one with the words, "surely we'll have the technology to deal with this". Another councillor considered that Oppenheim's offer – of "the opportunity for an appropriate community based organization in the Western Isles to take ownership of up to 25 per cent of the turbines on the Muaitheabhal windfarm", to use BMP's language (cited in CnES, 2005b: 8) – demonstrated that "this is an applicant very willing to work for the well-being of the community". No mention is made of the far greater profits that Oppenheim would reap from the venture. This was a matter of leasing the sites for turbines without cost, the community body itself having to purchase the turbines. As the land of the Eisgen Estate was not under crofting tenure, no land rent would be payable to a community.

At the meeting of the full council on 29 June, seven of the councillors opposed the motion to recommend the granting of planning permission for BMP, 22 voting in favour. Substantial opposition came from those living on the opposite side of Loch Seaforth in North Harris to where the turbines were to be sited, on account of their visual intrusion into the landscape. One of the Harris councillors spoke of the community council ballot taken on the subject in North Harris. With a record turnout of 70 per cent of the electorate, the results showed that 75 per cent of those voting were against the project. In a parallel ballot in South Harris, 67 per cent of those voting opposed the project, the turnout being 63.8 per cent. She concluded her contribution to the debate on 29 June by expressing concern about the lack of reflection of community views on the part of councillors. "I fear that we are not taking communities with us", she reflected. "We have to be leaders, yes, but not dictators". She was unconvinced that young people would return to the kinds of job needed on the wind farm or that the wind farm would suddenly stop youth from going away. They went away for reasons of diversity of career options.

Outwith the Comhairle debating chamber, it was clear that public opinion in Lewis was overwhelmingly against LWP's proposal. While the Muaitheabhal Wind Farm did not excite the same degree of opposition, it is evident that concern from people living in both Lewis and Harris was widespread. To turn first to LWP's proposed wind farm, the response from Ness Community Council (NCC), located on the Galson Estate in North Lewis, was particularly forceful (NCC, 2005, document attached to CnES, 2005c). At a meeting held on 20 June 2005, attended by over 100 people – an indication of the depth of feeling within the community and a concern that some of the leadership were out of touch with the majority of their constituents – the presentation of a report from a group of "concerned citizens" made it clear that they would brook no quarter from LWP. They insisted in unequivocal terms that the decision to reject the wind farm was "final, and not open to negotiation" (NCC, 2005:2, document attached to CnES, 2005c). The community was "totally opposed" to any suggestions about modifying the wind farm, to any resumption of crofters' common grazings, to giving quarrying consent to mineral resources on the estate and to the destruction of sites of considerable heritage value (NCC, 2005:2, document attached to CnES, 2005c). Recognizing the implications of LWP's proposal for the Galson Estate Trust's application to the Scottish Government to buy the estate, the report expressed concern that a decision on the wind farm by the government might be made while the community buyout was still under active consideration. The authors asked the government "to carefully judge the implications of announcing any decision that [might] jeopardise the buyout application" (NCC, 2005:1, document attached to CnES, 2005c). They went on to affirm their support of small-scale initiatives to generate electricity – from the wind, from waves and tides, or solar energy – as these would lead towards local self-sufficiency. And that, in turn, would "attract visitors, it would provide local employment, it would provide cheaper electricity for us all – everyone would benefit, and our moorlands, wildlife and habitats, and our unique heritage would remain intact for future generations to appreciate" (NCC, 2005:2, document attached to CnES, 2005c).[17]

The unprecedented number of individual objections to LWP sent to the Comhairle and to the Scottish Government provides further evidence of the degree to which the majority of elected officials at the Comhairle were at odds with public opinion. By the time that Banks's report was written, the Comhairle had received 1097 objections and the government 3656, of which, in both cases, the great majority were from the Isle of Lewis. The Comhairle received a further 68 submissions before the meetings at the end of June, all of them objecting to the wind farm

(CnES, 2005c). By the same time, the government had received a total of 4137 representations, of which only nine supported the wind farm (CnES, 2005c). The total number of representations received by the government reached 6131 in December 2005, 4573 (74.59 per cent) coming from residents on Lewis (*Stornoway Gazette*, 22 December 2005: 5).

With respect to BMP's Muaitheabhal Wind Farm, data contained in Banks's report indicate that 201 individual objections had been received by the Comhairle and 1349 by the Scottish Government, the vast majority coming from Lewis and Harris. In addition, the Comhairle had received two petitions from people opposing the wind farm. One has 60 signatures, almost exclusively from Harris (97 per cent of the responses), the remaining 3 per cent of the responses originating in Lewis. The other petition has 520 signatures, with the provenance of the signatories as follows: Harris 74 per cent; Lewis 23 per cent; Scotland 2 per cent; UK 1 per cent (CnES, 2005b: 360–361).

Beyond these measures of local opposition, highly visible resistance, directed particularly towards LWP, has come from Mòinteach gun Mhuileann/Moorland without Turbines (MWT), a Lewis-based non-governmental organization, and from the RSPB. Congruent with public opposition to the industrialization of rural landscapes elsewhere, which seeks to "protect" the landscape against corporate intrusion (Woods, 2003), MWT was established in 1994 with the objective of "protect[ing] the Lewis landscape, people and economy against inappropriate industrial developments", including wind farms and "the pylon-isation of the Highlands and Islands" (www.mwtlewis.org.uk, accessed 21 June 2011). In the discussion by van der Horst and Vermeylen (2008: 13), drawing on Michael Woods (2003), the group is part of "a new social movement" bent on preserving "the 'unspoiled' views and 'tranquillity' of the post-productive countryside as a new common". On Lewis, the group has sought to mobilize the public against wind farms through public meetings, press releases, the distribution of pamphlets to all households on Lewis and, more militantly, the torching of a model (30 foot) turbine beside Loch Grinnabhat, Benside, outside Stornoway, on 11 December 2004 (Fraser, 2004: 1).

In marked contrast, the RSPB's high-profile campaign against LWP's proposal centred on the publication in November 2005 of a series of maps that drew attention to the scale of the proposed operations (RSPB, 2005). Cartographically, the maps superimpose the proposed layout of LWP's wind farm on maps of the same scale of selected cities – Edinburgh, Glasgow and London – and their environs. Thus, for example, the proposed 234-turbine wind farm is shown as stretching from Edinburgh north to well beyond Kirkaldy and west to Dunfermline. The intent, according to the RSPB, was to demonstrate how the "overwhelming dominance" of the proposed wind farm would have "an unprecedented impact on endangered British birds and wildlife and the unique yet fragile habitat that supports them on the Hebridean island of Lewis" (quoted by MacSween, 2005: 2). Condemned as a "childish and patronising" publicity stunt, the Comhairle's Vice-Convener Angus Campbell sought to downplay the maps' significance by identifying the RSPB as an unelected single-interest group (MacSween, 2005: 2). Others criticized the organization for its alleged inconsistency in how it responded to proposals for wind farms in Shetland (*WHFP*, 26 January 2007: 2, Editorial: 13)[18] and on its own reserves (*WHFP*, 2 January 2009: 3).

On 21 April 2008, the government turned down Lewis Wind Power's application for planning permission, despite the fact that the earlier proposal had been modified in accordance with at least some of the conditions imposed by the Comhairle in 2005, including a reduction in the number of turbines from 234 to 181. Those situated close to settlements and those considered significant in minimizing environmental costs had been withdrawn (*WHFP*, 16 February 2007: 5; MacInnes, 2007: 3; *WHFP*, 23 February 2007: 7). In reaching his decision, Jim Mather, Energy Minister, indicated that he had taken into account the Comhairle's position, the (by then) 10 924 letters of objection, and the 98 letters of support, but stated that the proposal was refused on the grounds of its "significant adverse impacts on the Lewis Peatlands Special Protection Area" and thus its non-compliance with European legislation (cited in *Stornoway Gazette*, 21 April 2008: 11–12). He did not, however, rule out all future initiatives, and announced that the Scottish Government was commissioning a study to identify renewable energy potential in the Outer Hebrides. The resulting report by the Halcrow Group Ltd (2009: 105) identifies an area south west of Stornoway, well outwith the SPA, as providing the greatest potential for a wind farm with up to 150 MW generating capacity. With the support of the Stornoway Trust, LWP is taking forward a proposal for a 42 turbine farm to be located on a site recommended by Halcrow (*WHFP*, 2 July 2010: 3).

In January 2010, the Scottish Government announced its approval of the Muaitheabhal Wind Farm with a reduced number of turbines, 33, with a generating capacity of 118 MW, compared with 133 turbines and a generating capacity of 399 MW in BMP's initial proposal (*WHFP*, 22 January 2010: 3). The initial reduction – from 133 to 53 turbines – took place in April 2006 in response to the number of objections that the proposal had sustained (MacInnes, 2009b: 3).[19] This amended proposal was then the subject of a public local inquiry, held in Stornoway in

May 2008. The Reporter's conclusions, made public in August 2009, were that the conditions detailed in Paragraph 25 of NPPG 14 were not met (MacInnes, 2009c: 1). Namely, the integrity of the South Lewis, Harris and North Uist NSA would be compromised and the adverse effects would not be outweighed by social and economic benefits of "national" significance, except insofar as the wind farm might potentially contribute to the construction of an interconnector (MacInnes, 2009c: 1). By this time, BMP had informed the government that it was revising downwards to 39 the number of turbines (MacInnes, 2009b: 3), from which number the government removed a further six turbines on the grounds of reducing the visual impact of the scheme, thereby securing SNH's approval (*WHFP*, 22 January 2010: 3; MacInnes, 2010: 1). Under Section 75 of the final agreement – as a condition for planning permission – Oppenheim has agreed to sub-lease sites for four turbines to the Muaitheabhal Community Wind Farm Trust (MCWFT) and to pay 1 per cent of the income to the MCWFT and 0.5 per cent to the Western Isles Development Trust. Local councillor, Annie MacDonald, welcomed the decision. "I firmly believe", she said, "that this is not a hand out but a 'hand up' which will allow the community to shape its own destiny" (cited by MacInnes, 2010: 1).

While not denying that the community may reap some potential gains from BMP's Muaitheabhal Wind Farm, it needs to be recognized not only that the distribution of profits will disproportionately favour the land owner, but that this process will in turn reinforce the prevailing pattern of land ownership and, specifically, the ownership model of property. The leasing of four sites for wind turbines to the community may complicate the ownership model of property, but Oppenheim's offer simultaneously acts to legitimate the accumulation of vast wealth by an individual and the normalization of this model of property. Community is recruited in the pursuit of privatization – of the wind and of the land, of nature and of property. With respect to property, as I indicated earlier, the estate is not under crofting tenure and therefore its residents do not have an absolute right to purchase the land under Part 3 of the Land Reform Act. Under Part 2 of the Act, they do have the right, as a community, to register an interest in the land and to exercise a pre-emptive right conditional on the owner's decision to sell. In a situation where commodifying the wind can bring such rewards, it is hardly in that individual's interest to divest her- or himself of rights to the land. Even were this the case, the dramatic increase in the value of land that accompanies the granting of planning consent all but prohibits community purchase. In these circumstances, the community becomes complicit in a process that pushes their rights as detailed under Part 2 of the Land Reform Act even further out of reach and fuels calls for more robust land reform legislation (Wightman, 2010, 2011).

To extend the argument, the co-option of community is a ruse allowing the norms of commodification of nature to proceed. Privatization of property proceeds through the privatization of a commons, the wind. To use Katz's (1998: 46) phrase once more, through the wind, nature becomes "an accumulation strategy for capital". By discursively separating a claim to nature from a claim to property – by depoliticizing the enclosure of the wind – the planning process is complicit in this process.

Claiming property, capturing nature

I have suggested above that, in the case of the Eisgen Estate, the process of capturing the wind proceeds according to the norms of commodification of nature, a process which reinforces the ownership model of property and in which residents of that estate are complicit. I turn now to show, through the examples of the Galson Estate and Pàirc, how crofting community trusts disrupt these norms through their struggle to exert their rights to the land. These trusts were the first to submit applications to the Scottish Ministers under Part 3 of the Land Reform Act – "hostile bids" – on account of the land owners' unwillingness to sell. Both are also located on land where private land owners were seeking planning permission for large-scale wind farms, a matter that, if resolved in the applicants' favour, would lead to soaring land values. In addition, both community trusts had to negotiate for the land where rights had become further complicated by the owners' establishment of an interposed lease, a legal device designed to ensure that, were the land to be sold against their wishes, sellers would retain "development" rights.[20] In both cases, the sellers stood to reap substantial profits should the proposed wind farms go ahead. The stories provide clear evidence of communities' attempts to resist dispossession through challenging the norms of property and nature.

I turn first to the experiences of the Urras Oighreachd Ghabhsainn/ Galson Estate Trust, which celebrated its purchase of the 56 000 acre Galson Estate in North Lewis, almost all of which is under crofting tenure, on 12 January 2007. The buyout followed four years of community consultation and then, with the refusal of the land owners, Galson Estate Limited,[21] to sell the land, an application to the Scottish Ministers under Part 3 of the Land Reform Act. Interest in a community buyout of the estate preceded both the Land Reform Act and Lewis Wind Power's proposal for a wind farm, but the struggle for the land quickly became entangled in the struggle to capture the wind. As James MacDonald, chair of the Galson Estate Crofting Community Land Ownership Steering Group, explained in a public meeting in September 2004, rights in property were closely intertwined with rights to the wind:

> As long as the estate is out of our hands, there is nothing we can do to stop the place being sold to the highest bidder. If this wind farm goes ahead Galson Estate will suddenly be worth an awful lot of money, making it an attractive purchase for an international company – who knows, even for Amec ... If we are to have any control in the future development of this estate we must stand together now and make the Galson Estate Trust a worthy vehicle to drive our community's aspirations forward (cited in *WHFP*, 10 September 2004: 2).

However, the route in pursuit of community ownership was not straight-forward. At one point, tension emerged between the steering group (and later the board of Galson Estate Trust), which was seen by many as interested in the income the wind farm would bring, and a membership which was increasingly mobilizing in opposition to the proposed wind farm (interview, member of the Galson Trust's Board, 15 December 2008).[22] A ballot held in June 2005 showed that 83 per cent of the residents on the Galson Estate opposed the wind farm and the actions of the land owners, who had agreed to the development without consulting the community (*Stornoway Gazette*, 23 June 2005: 3). The owners, named as Galson Estate Limited had, in fact, signed a lease agreement with Lewis Wind Power through the Galson Energy Company, a company to which development rights to the land of the estate were hived off (*Stornoway Gazette*, 23 June 2005: 3). This was, in effect, an interposed lease which would allow the owners to benefit from any developments to the land were the land to be sold against their wishes.

A subsequent ballot showed substantial community support for a "hostile bid" for the land under the terms of Part 3 of the Land Reform Act in the face of the owners' unwillingness to enter into negotiations: 85 per cent of those voting were in favour, with 72 per cent of those eligible casting a vote (http://urras.org/wiki/TheTrust, accessed 17 November 2008). However, the process of following this route was not easy. Part 3 of the Act (Section 73) contains precise conditions that a community must meet, including the requirement that the land in question had to be mapped accurately. Each croft boundary as well as the extent of the common grazings had to be identified, as well as all "sewers, pipes, lines, watercourses or other conduits and fences, dykes, ditches or other boundaries in or on the land" (Section 73 [5][a] and [b] [i], [ii]) – an onerous and time-consuming task. As one informant commented, some records were missing, others were inaccurate and mapping was thus "a very complex business that could have gone on for years" (interview, 15 December 2008). Because it was recognized that time was of the essence, given the impending planning decision regarding Lewis Wind Power's proposal, the Galson Trust decided to apply initially only for purchase of the common grazings which were relatively straightforward to map, leaving until a later date the purchase of the inbye land (*WHFP*, 3 June 2005: 1). In the event, recognizing that the community trust was serious in its pursuit of a buyout and that, in the absence of a court ruling confirming the legality of an interposed lease, they risked losing everything – the land and development assets – the owners took the safer route and agreed to negotiate "amicably" (interview, member of the Galson Trust, 15 December 2008). Once the Galson Trust withdrew their application to the Scottish Ministers, mapping of the inbye land became unnecessary and the estate was considered as a whole for the purpose of sale.

The terms of the purchase were complex.[23] Unable at the time to mount a legal case against the interposed lease – and the former owners' agreement with Lewis Wind Power – the Galson Trust bought the lease together with the land. It thereby took on Galson Energy, later converting it legally to a trading company, as required by the legislation pertaining to a company limited by guarantee (interview, Galson Trust member, 15 December 2008). A condition of the sale was that, were the wind farm proposed by Lewis Wind Power to be granted planning permission, the previous owners would receive 17.5 per cent of what they would have received had they remained landlords; 82.5 per cent would go to the community, an arrangement that would last for 25 years (*WHFP*, 19 January 2007: 7).[24] A further condition stipulated that, if the trust allowed other private wind initiatives within seven years of the date of purchase, the sellers would receive 10 per cent of the trust's share of the income for a period of 15 years (*Stornoway Gazette*, 6 October 2005: 3). In the case of a community wind farm, the sellers would be entitled to purchase 15 per cent equity in the development (*Stornoway Gazette*, 6 October 2005: 3). As discussed earlier, the Scottish Government's rejection of Lewis Wind Power's proposal has laid the controversial project to rest. "No-one can ever do that to us ever again", said one interviewee, "because we own the land" (interview, Galson Trust member, 15 December 2008). The Galson Trust is now pursuing its own proposal for a three-turbine wind farm, which is expected to produce an annual income of £150 000 (*WHFP*, 12 January 2007: 3).

In the case of Galson, the land reform legislation may not have been tested in the courts of law but it undoubtedly played a key part in re-balancing the relationship between private land owner and crofting tenants/residents on an estate. As such, discursively, it became a means through which the norms of commodification were disrupted and a counterhegemonic, community-centric, ethic *vis à vis* the exploitation of the wind created. However, this process was complex. The Galson Trust decided against testing the legality of an interposed lease in court, action that Pàirc Trust was required to take, as I discuss shortly. Instead, it bought the lease together with the land, renegotiating its terms with the sellers. The boundary between private seller and community trust buyer *vis à vis* the land was thereby complicated, the seller retaining some development rights to the land even as the land entered community ownership. In effect, then, while community ownership of the land has clearly subverted the norms of private property, the creation of a community ethic through any future commodification of the wind will be mediated, in part, through a subscription to private interests and a private ethic – whether the enclosure of the wind occurs through a community owned wind farm or through a corporate or private initiative – at least for a period of time. There is a complex interweaving of community and private interests rather than an outright rejection of neoliberal norms of enclosure.

With respect to Pàirc Trust, community purchase of the land is in process. As indicated earlier (Chapter 2), the Scottish Ministers gave their assent to the community's application to purchase the common grazings of the Pàirc Estate and an associated interposed lease under Part 3 of the Land Reform Act in a landmark decision in March 2011. For the first time the legislation had been put to the test and the community had emerged the victor, or so it appeared at the time. In accordance with the legislation, the trust was given six months to raise funds for the purchase, following a district valuer's assessment of the purchase price. That process was subsequently put on hold as a result of the land owner's decision to take the Scottish Government to court on several grounds. These included the grounds that the Land Reform Act contravened the European Convention on Human Rights. The case was heard in the Stornoway Sheriff Court on 23 June 2011, and the matter of human rights pertaining to European law was referred to the Inner House of the Court of Session in Edinburgh. A date for the hearing has yet to be set. In the meantime, discussions between the trust and the land owner continue (communication, Pàirc Trust spokesperson, 2 November 2011).

From its inception in December 2003 as a company limited by guarantee with a remit to purchase the 26 790 acre (10 846 ha) Pàirc Estate, almost all of which is under crofting tenure, Pàirc Trust faced the opposition of the absentee landlord, Barry Lomas, an accountant who resides in

Leamington Spa, England, and who controls Pàirc Crofters Ltd, the land owning body. At the time of the trust's formation, Lomas had just signed a "Heads of Agreement" contract with Scottish and Southern Energy, which had proposed a 125-turbine wind farm on the estate (*WHFP*, 26 December 2003: 3). The new trust had requested that any such discussions be delayed for one month until their elections to name the new directors had been held, as it would be this group of people who would have the community's mandate to take their position forward in discussions with the land owner and Scottish and Southern (*WHFP*, 26 December 2003: 3). In May 2005, unable to make any headway in negotiations with the owner, Pàirc Trust submitted an application to the Scottish Ministers to purchase the common grazings under the terms of Part 3 of the Land Reform Act 2003 – in other words, as a hostile bid. As in the case of Galson, the inbye land was excluded from the application on the grounds that the mapping requirements were prohibitive. In their application, the trust questioned the legality of the interposed lease set up by Pàirc Crofters Ltd in 2004 and, as such, made no offer to purchase the lease. The lease had been established, as was the case in Galson, in anticipation of a community trust application to buy the estate and with a view to gaining from any wind farm development of the land. Pàirc Crofters had set up a wholly owned subsidiary company, Pàirc Renewables Ltd, to which development rights were leased for a period of 75 years (interview, Pàirc Trust spokesperson, 3 July 2008).[25] In the event of a community buyout, as could have been the case with Galson Energy, through an interposed lease, Pàirc Renewables would, to all intents and purposes, control the estate, despite the fact that the Pàirc Trust would be the owners, having bought the land from Pàirc Crofters (interview, Pàirc Trust spokesperson, 3 July 2008).

It is important to note that Pàirc Trust, unlike the Galson Trust, was not disputing the location of the wind farm on the estate. Indeed, Donald MacDonald, the spokesperson for what was then the Pàirc Liaison Group (the precursor to the Pàirc Trust),[26] indicated in August 2003 that the "very large revenue stream" that would be generated by the wind farm would provide the financial means for the community to buy the estate (cited in *WHFP*, 22 August 2003: 9). Early calculations from Scottish and Southern suggested that £400000 would be generated annually, an amount that would be divided equally between the landlord and crofters until such time as a buyout, when the total sum would accrue to the community (*WHFP*, 22 Aug 2003: 9).[27]

Despite the expected financial rewards of the venture, the matter had, in fact, caused considerable dissent within the community. At one point, a member of the Pàirc Protection Group, an organization set up in opposition to the Liaison Group, was warned that she could face legal action if she did not retract some of her statements, which she subsequently did (*WHFP*, 22 August 2003: 9). For Donald MacDonald, the viciousness of the verbal attacks had the opposite effect to that intended. The protestors had fostered the creation of "a great spirit of unity" (cited in *WHFP*, 29 August 2003: 2). The ballot held on 29 November 2004 confirmed his analysis. The results demonstrated that the majority of residents were solidly in support of the community buyout. Of a turnout of almost 70 per cent of those eligible to vote, 222 (87.4 per cent) were in favour of proceeding with a hostile buyout, while a mere 32 (12.6 per cent) opposed the proposal (*WHFP*, 3 December 2004: 1). Among crofters, with a turnout of 77.9 per cent, 99 (85 per cent) were in support – a figure well above the simple majority required by the land reform legislation; only 17 (14.7 per cent) were opposed (*WHFP*, 3 December 2004: 1).

As there had been no court ruling on the matter of the legality of an interposed lease with respect to land under crofting tenure, the government required Pàirc Trust to seek clarification through the Scottish Land Court. The court heard the case in June 2007 and issued their finding that the interposed lease was valid in August 2007 (interview, Pàirc Trust spokesperson, 3 July 2008). By this time, however, the Scottish Parliament had effectively closed the loophole in Part 3 of the land reform legislation by including, in the Crofting Reform (Scotland) Act 2007, a provision that allowed crofting communities to purchase the interests of the holder of the interposed lease (interview, Pàirc Trust spokesperson, 3 July 2008). Pàirc Trust was thus now in a position to purchase the interests of Pàirc Crofters Ltd and Pàirc Renewables, which position they thought they were in at the time the trust was formed (interview, Pàirc Trust spokesperson, 3 July 2008).

Subsequent to the court ruling, the government indicated that the trust should submit new applications – one for the land and one for the interposed lease – which meant that a further ballot of the community was needed. This ballot, carried out independently through the auspices of the Comhairle in December 2009, showed substantial support for both a buyout of the estate and the purchase of the interposed lease: 69 per cent of those who voted were in favour, the turnout being 75 per cent of eligible voters; 72 per cent of crofters voted in favour, again exceeding by a wide margin the majority required under the Land Reform Act (MacInnes, 2009d: 4). And, on that basis, frustrated by years of negotiations with the landlord, who had repeatedly insisted that he wanted to reach an "amicable" agreement while threatening legal action, Pàirc

Trust renewed their pursuit of a hostile bid in March 2010 (*WHFP*, 5 March 2010: 3). Lomas's efforts at negotiation were seen by the trust as "delaying tactics", buying time while the value of the interposed lease grew in anticipation of planning permission for a 26 turbine wind farm by Scottish and Southern Energy, to whom Pàirc Renewables had granted a sub-lease of part of the estate (Pàirc Trust, 2009).

For those supporting the buyout, purchasing the estate is about much more than gaining income from a wind farm. Angus McDowall, Chair of Pàirc Trust, insists that the buyout is "overwhelmingly in the public interest" and the "key to reversing a century of decline" (McDowall, cited in *WHFP*, 5 March 2010: 3). "Land struggle and land reform have a long history in Pàirc", he said; "[e]ver since the clearances of the first half of the nineteenth century, the area has been at the forefront of efforts to right historic wrongs" (McDowall, cited in *WHFP*, 5 March 2010: 3). He referred to the Pàirc Deer Raid about which Murdo Morrison had spoken in the ceremonies celebrating the North Harris Trust's purchase of that estate (Chapter 2), and the more recent raids of the 1920s, which led to the establishment of the crofting township of Orinsay. More recent problems, he noted, centred on a continuing trend of population decline – the present day population being under a quarter of the figure of 100 years ago (McDowall, cited in *WHFP*, 5 March 2010: 3).[28] "We believe", he emphasized, " that community ownership of the land, together with acquiring the interposed lease set up by the landlord in 2004, is the key to halting and reversing this debilitating trend" (McDowall, cited in *WHFP*, 5 March 2010: 3). Their plans for "sustainable development", he said, included not only renewable energy projects, which would provide jobs, but also the building of social housing.

Part 3 of the Land Reform Act has now finally been put to the test and a legal precedent set. The bastions of the ownership model of property have been well shaken. If, with private tenure, nature – through the wind – becomes a means through which capital is accumulated (see Katz, 1998), community land ownership opens up the process of commodification of nature to different, more complex, configurations. While the details of the agreement with Scottish and Southern Energy are not yet in the public domain, two observations may be made. On the one hand, through the commoning of property, nature may still become an accumulation strategy, but the process of commodification is mediated by a collective ethic. The enclosure of the wind works to strengthen a collective "becoming". Income from the wind farm holds the possibility of supporting more socially just community objectives than would have been the case with private ownership. There is here an inversion of processes of neoliberalization. On the other hand, as the turbines in the proposed wind farm will be owned by a corporation accountable to its shareholders, nature will at the same time be conscripted more directly into norms of commodification – and processes of neoliberalization – than would be the case were the turbines owned by the community.

Claiming nature, working the wind

I extend this discussion about troubling the norms of property and nature by considering how a land owning community trust aiming to build a small-scale wind farm countered the opposition of the statutory conservation authority, Scottish Natural Heritage. It did so, I argue, by troubling the idea of nature as external to social process, through which binary the conservation authority on this occasion legitimated its position. The community's production of nature instates a new politics that points in the direction of a more socially just enclosure of this commons than that prefigured by the conservation body.

While the story of the North Harris Trust's efforts to build a three turbine wind farm at Monan ended in December 2009 with the community trading company's decision not to proceed, its experience in progressing this initiative through the planning process – specifically, in dealing with the objections raised by SNH – provides a window into how a community's plan to capture of the wind disturbs norms through which nature is commodified.[29] My focus is on the ontologically different productions of nature of SNH and of the NHT. I draw on documents produced for the Comhairle's Environmental Services Committee, principally those authored by SNH and the NHT, associated policy documents, and interviews with NHT directors and staff.

In common with such initiatives of other land owning communities, the NHT's ambitions for a small-scale, community owned wind farm centred on the reversal of decades of social, cultural and economic decline. The wind, captured and commodified through its sale to the national grid, was to be one of the primary means whereby a community economy could be realized. To this end, through its wholly owned subsidiary, the North Harris Trading Company (NHTC),[30] with logistical and financial assistance from what was then called the Highlands and Islands Community Energy Company (HICEC), the NHTC undertook the necessary tasks prior to submitting an application for planning consent to the Comhairle.[31] The site at Monan, 60 m north west of the Ceann an Ora quarry near Aird Asaig, had been chosen following a detailed energy audit carried out for the trading company by West Coast Energy Ltd in 2004.[32]

From the start, the NHT and its trading wing had involved the community – through public displays of the options and open discussion. They had also initiated meetings with the various consultees required by the planning process. These included the Comhairle, SNH, Historic Scotland and the Scottish Environmental Protection Agency. Following both sets of meetings, and a ballot that showed that 93 per cent of those residents responding supported the project (interview, NHT Development Manager, 5 December 2006), the NHTC submitted its application for planning permission for a three turbine wind farm to the Comhairle in May 2006. After a substantial delay, the NHT's application was finally discussed at a meeting of the Environmental Services Committee on 22 March 2007. Agreeing unanimously with the recommendation of the Comhairle's planners that the proposal be approved, members of this committee took the further step of acting as a "committee of the whole" – thus obviating the need for deliberation by the council in its entirety – and sending the proposal to the government without further debate. This action was deemed appropriate given the short time line before the elections to the Scottish Parliament, which would delay proceedings. The action was necessitated by the sustained objections of SNH to the wind farm.[33] The government's subsequent call for a public local inquiry was only cancelled after SNH withdrew its objection in early September 2007, following substantial local outcry (Editorial, *WHFP*, 20 July 2007: 15; D. Mackay, 2007: 13; K. Mackay, 2007: 15; Mackenzie, 2007: 15). Planning consent was finally granted by the Comhairle in February 2008.

SNH's objection to the proposal came as a surprise to the NHT. Initially, SNH objected to the proposal on the grounds that the wind farm could potentially have a negative effect on golden eagles, one of the qualifying species for the adjacent North Harris Mountains SPA, an objection it later withdrew, and on the grounds of compromising landscape and visual amenity. Situated within the South Lewis, Isle of Harris and North Uist NSA, the turbines would be within sight of An Cliseam, the highest hill in the Outer Hebrides, from a small section of the main road between Tairbeart and Stornoway. Although situated outwith the SAC, the SPA and the SSSI within whose boundaries An Cliseam is located, the wind turbines, or so it was argued, would disturb the sense of "wildness" conjured by An Cliseam (Photograph 4.1).

This was not the first time SNH had been at the centre of controversy with respect to a community's ambitions to generate electricity from a renewable source. The organization's opposition to the siting of a small hydro-electric scheme on Loch Poll, proposed by the Assynt Crofters' Trust shortly after its purchase of the North Assynt Estate in 1993, is well known.

Photograph 4.1   An Cliseam.
Source: author.

Here, SNH's objection centred on the potential disturbance to a pair of rare black-throated divers, whose nesting pattern was at risk (MacKenzie, 2003).

SNH's politics insofar as the proposed wind farm at Monan is concerned are made clear in two letters written by David MacLennan, Area Manager for the Western Isles and Rum, dated 1 November 2006 and 12 December 2006 and addressed to the Comhairle. In these letters, MacLennan draws on both the Natural Heritage (Scotland) Act 1991 and on National

Planning Policy Guideline 6, *Renewable Energy*, to legitimate SNH's objections to the proposal on the grounds of landscape and visual amenity. National Scenic Areas, wrote MacLennan in his first letter to the Comhairle,

> are designated for their scenery of unsurpassed attractiveness and are considered to be of national importance. NSAs are particularly sensitive to inappropriately designed and sited development which would adversely affect the qualities for which they are designated. The description of the NSA in Scotland's Scenic Heritage (1978) describes the qualities of North Harris as distinct from other parts of the NSA. The mountainous character is considered to be a key quality of this part of the NSA (in CnES, 2007b: 141).

While MacLennan acknowledged that the proposed location was already scenically compromised – it lies close to the main road between Tairbeart and Stornoway, the Ceann an Ora quarry, numerous pylons, powerlines and a "lattice style" communications mast – he drew on the *Landscape and Visual Impact Assessment (LVIA)* (Horner and MacLennan, 2006) carried out for the NHT to conclude that "there will be significant impacts upon the character and qualities of this part of the NSA and that there would be significant impacts upon visual amenity within this part of the NSA" and therefore that SNH objected (CnES, 2007b: 141). What he omitted to draw attention to – highlighted in the report of the Director of Sustainable Communities to the Environmental Services Committee, meeting in March 2007 – was that the *LVIA* nowhere suggested that "the overall integrity of the area" would be compromised by the wind farm. The *LVIA* noted that the effects on landscape and visual amenity would be local. This issue, the director's report made clear, is the key test for acceptability in assessing whether the proposal should be supported or not, as indicated in Paragraph 25, NPPG 14 (see the discussion of Beinn Mhor Power above). Only if "the overall integrity of the designated area was compromised", the report stated, would it "have to be shown that [the adverse effects on the qualities for which the area has been designated] would have to be outweighed by the national benefits" (CnES, 2007b: 128–129; see Mackenzie, 2009). The second document by landscape architects Horner and MacLennan, dated February 2007, made a blunt assessment of the problem:

> Unfortunately, SNH has not applied this test [of whether the overall integrity is affected, as indicated in NPPG 14], although they state in their *Landscape Policy Framework* [SNH, 2005a] that it is the basis on which they will judge acceptability within NSAs (Horner and MacLennan, 2007: 1).

They conclude that,

> if the test for acceptability within NPPG 14 is correctly applied, the land-
> scape and visual impacts of the proposed development within the Lewis,
> Harris and North Uist NSA would be judged as being localized and that
> they would not affect the overall integrity of the NSA (Horner and
> MacLennan, 2007: 5).

In justification of SNH's position, MacLennan drew on Section 3 of
the Natural Heritage Act. SNH, he wrote in his first letter to the
Comhairle, had a "balancing duty" – to take account of not only ecological and environmental matters, but also the "needs" of agriculture,
fisheries and forestry, "social and economic development" and "sites
and landscapes of archaeological and historical interest" and the
"interests" of owners and occupiers of land and local communities
(in CnES, 2007b: 142). With respect to Monan, MacLennan considered that:

> The need for economic development in North Harris, the interests of the
> North Harris Trust and the interests of the local community may all be
> taken into account as part of our balancing duty in any assessment of this
> application. In this case SNH must consider whether those interests out-
> weigh our concerns over the significant landscape and visual impact upon
> the National Scenic Area (in CnES, 2007b: 142).

He continued,

> NPPG 6 states that in NSAs *"renewable energy projects should only be
> permitted where it can be demonstrated that the objectives of the designa-
> tion will not be compromised or any significant effects on the qualities for
> which the area has been designated are clearly outweighed by social and
> economic benefits of national importance* (in CnES, 2007b: 142, emphasis
> in original).

Thus, he concluded,

> So if the adverse impact on the NSA is to be outweighed, the social and
> economic benefits must be of national importance. SNH is of the view that
> the current proposal cannot be considered to be of national importance.
> We have therefore concluded that we have insufficient basis for using the
> balancing duty so as to override our concerns about NSA impacts in this
> instance (in CnES, 2007b: 142).

What is striking is MacLennan's omission of SNH's policy statement, *Strategic Locational Guidance for Onshore Wind Farms in Respect of the Natural Heritage*, published in May 2005 (SNH, 2005b). With specific reference to National Scenic Areas, SNH distinguishes between "commercial wind farms" and those that are community owned. "Proposals of a modest kind", the document states, should "in addition to turbines attached to 'individual properties' be given especially sympathetic consideration where they contribute to the sustainability of an isolated community such as an island" (SNH, 2005b, Section 1.2).

Justifying his position further, MacLennan noted that the proposed wind farm was "on the southern periphery of an area of search for wild land" (in CnES, 2007b: 142). As specified in SNH's *Guidelines on the Environmental Impacts of Windfarms and Small Scale Hydroelectric Schemes* (SNH, 2001: 45),

[The wild qualities of a landscape] depend on a strong sense of sanctuary and solitude, a high degree of naturalness in the setting, a freedom from human activity or influence, the absence of any built structures and a sense of awe or, for some, a sense of challenge. ... Wildness is a quality which people can find in many different parts of the Scottish landscape, sometimes close to settlements, but normally best expressed in remote upland and coastal areas, especially those which lie beyond the existing network of public and private roads.

The assessment by Horner and MacLennan (2006: 93), to which MacLennan did not refer, considered that the turbines at Monan "would result in low magnitude of impact and minor significance of impact on wild land and wildness qualities".

Through an invocation of the wild – isolated from the currents of historical and contemporary process – and of a "balancing duty" assessed against "national importance", SNH attached particular meanings to the land as "landscape" and as "visual amenity". It conjured "nature" as external to the workings of the social, a production that, to follow Toogood's (2003: 153) line of reasoning, "reflects conservation's colonial legacy", a legacy that, "at the very least", facilitates the "side-step[ping of] issues of power and justice connected with the land". "At the worst", he suggests, it "reinforce[s] the construction of the Highlands most closely associated with the large estates and their practices" (Toogood, 2003: 153). This production of nature "writes against" the political possibilities opened up through community land ownership.

For the North Harris Trust, responding to SNH's representations, this was a land whose workings were integrally entwined with political

156   PLACES OF POSSIBILITY

process – locally, regionally (the Hebrides), and nationally – not somehow immured from it. The proposed wind farm, however small, was key not only to the "reinvigorat[ion]" of "all aspects of North Harris life – social, economic, natural and cultural", but also to the success, nationally, of the land reform process, or so the NHT maintained (Cameron to the Comhairle, letter dated 6 November 2006: 6). David Cameron, NHT Director leading the initiative, made the case as follows.

Carefully positioning the NHT's counter-argument with reference, first, to two policy documents (SNH Policy Summary 00/15 and SNH Policy Statement 05/01, "SNH's Landscape Policy Framework") that emphasize the need to balance "scenic interest" and "landscape change" with community need and, second, to the "serious social and economic situation" in Harris and elsewhere in the Outer Hebrides, Cameron identifies two critical factors underlying the crisis: a lack of capital and a lack of confidence (Cameron to the Comhairle, letter dated 6 November 2006: 6). It is these, he suggests, that lie behind population loss (from about 4000 on Harris in 1951 to about 2000 in 1991), school closures, a housing shortage and a level of substandard housing above the national average, limited job opportunities and a decline in active crofting (Cameron to the Comhairle, letter dated 6 November 2006: 4–5). The community's purchase and management of the North Harris Estate, he stated, had begun to redress the "'confidence' problem" (Cameron to the Comhairle, letter dated 6 November 2006: 5). But, he emphasized,

> without the capital, there will be no further progress. There will be abso-lutely no chance to reverse the North Harris situation. There will be no chance to become a sustainable community. … The community have not previously had an opportunity like this [i.e. the wind farm] and there is nothing on the horizon to suggest that we will ever get the chance again (Cameron to the Comhairle, letter dated 6 November 2006: 5).[34]

As great as the significance of the wind farm at Monan might be locally, Cameron continued, it was of additional importance at the Hebridean scale. A number of communities had been encouraged by the North Harris initiative – Colonsay, Jura and Islay being examples – not least through the NHT's hosting, with the financial support of HICEC, of a conference in April 2006 focused on renewable energy. Entitled *Community Energy: Leading from the Edge*, the conference was about the sharing of expertise, drawn from communities themselves as well as from HICEC and others, as well as being about the building of links among community groups aspiring to build renewable energy projects.

The Monan project, Cameron wrote, "is a catalyst for others" (Cameron to the Comhairle, letter dated 6 November 2006: 5).

Compellingly, as a counter to SNH's attempt to capture the "national" high ground in its objection to the proposed wind farm, Cameron legitimated the initiative through specific reference to the Land Reform Act 2003. Community ownership, he insisted, "and also the sponsorship and development of community ownership projects, which provide a sustainable future for these communities, are themes of direct national importance in Scotland" (Cameron to the Comhairle, letter dated 6 November 2006: 6). He continued,

> The project is of national importance in that it allows a community trust to become a significant agent of regeneration and gives it the tools to carry out that "sustainable development" identified in the Land Reform Act. It also acts as an emblem of community empowerment and is of national importance in that it also gives confidence and inspiration to other communities thinking of going down the community energy route (Cameron to the Comhairle, letter dated 6 November 2006: 6).

In other words, there is a direct link between the wind farm at Monan and the national land reform agenda. Without such initiatives, the purpose for which the Land Reform Act was legislated would be undermined.

Cameron furthers the NHT's claim to national significance by direct engagement with the arguments presented by SNH in MacLennan's letters. With specific reference to NPPG 14, and drawing on the *LVIA* of Horner and Maclennan (2006), he insisted, first, that the impact on landscape and visual amenity would be localized and would in no way compromise the integrity of the NSA. Second, referring to Paragraph 24 of NPPG 6, Cameron claimed that the initiative was of national significance. To quote:

> Small schemes will provide a limited but valuable contribution to renewables output and to energy requirements both locally and *nationally*. Planning authorities should not reject a proposal simply because the level of output is small (NPPG 6, quoted in Cameron to Comhairle, 6 November 2006: 4, Cameron's emphasis).

He extended the argument in his second letter to the Comhairle:

> If you accept that the well-being of a region of Scotland such as Harris is of national importance, then one project, Monan, which has the potential to make a significant demographic contribution and reverse the current

decline, is also of national importance. ... If the project goes ahead the Trust will have an average of £125 000 per year to invest in the area. Without it, the Trust will return to the role of previous landlords in collecting rents from a declining community (Cameron, letter to the Comhairle, 22 February 2007: 1).[35]

The NHT's case for the wind farm is presented as a claim to a particular place, linked to the struggles of other Hebridean communities which are similarly struggling to turn around historical and contemporary processes of dispossession, and integral to the achievement of a national agenda, whether this concerns the possibilities promised through land reform or the attainment of renewable energy targets. It is a case that recognizes the embeddedness of the process of capturing the wind in politics – of the land and of nature – at scales from the local to the national and the global. "Nature", as produced in the NHT's arguments presented to the Comhairle, is not pitted against the economic, the social, the cultural, the political. It is produced as part and parcel of the land a-working in the interest of a collectivity's search for a more sustainable future. Inasmuch as this production of nature troubles SNH's insistence (in this instance) on the binary social/nature, it also disturbs the norms of commodification. Sold as a commodity – electricity – to the national grid, working the wind is evidently "complicit with capitalism" (Watts, 2004: 197). But, mediated through a community ethic rather than corporate or individual interest, the wind becomes a means through which a community can exercise collective rights. In this sense, the enclosure of the wind – the production of nature as a commodity – becomes a vehicle through which rights to the land are affirmed and a community economy brought into being.[36]

Towards a more social nature – "through and with" the wind

It is difficult, writes Michael Watts (2004: 195) in his incisive critique of petro-capitalism in the Niger Delta, not to "see" communities "through and with" oil. However, unlike oil, and similar to water and solar power, wind is not only a renewable resource, it is also a divisible resource. It can be captured by small-scale, land owning, communities. Commodified, the wind can provide the means through which communities may configure an alternative, more generous, imaginary of the future than that possible through corporate or individual control. It can become the means through which people reposition themselves *vis à vis* both property and an economy that, to follow Gibson-Graham (2006: 87), is more "place-attached", community owned, led and controlled, has "communal appropriation and distribution of surplus", and is "environmentally sustainable", "small scale", "locally self-reliant" "culturally distinctive" and "socially embedded".

As yet, it is primarily from Gigha that well-documented evidence of the significance of community ownership of a wind farm and of its contribution to the creation of a community economy is found. Achievements have included a housing improvement and building programme, new business premises and businesses, and a growth in population of close to 50 per cent since the islanders took over ownership. There were 98 residents when the Gigha Heritage Trust took over; the current population is over 150 (Hunter, 2012: 133). Enrolment in the primary school increased from 6 in 2002 to 22 in 2011 (Hunter, 2012: 133). The housing programme was seen as vital to this regeneration, as 75 per cent of the houses were "below tolerable repair and the rest were in a serious state of disrepair" (Morrison, 2006: 10). In addition to repair and refurbishment of the housing stock, 18 energy-efficient homes have been built at a cost of £3 million, funded by the Scottish Government (Morrison, 2006: 10). There are now plans to build a fourth turbine (Sweeney, 2009: 26). By the community's direction of the surplus generated from the sale of electricity towards the building of a community economy rather than its distribution as dividends to globally situated shareholders or into the bank accounts of wealthy individuals, the turbines are a means through which people are in the process of renegotiating their subject positions through the creation a community economy.

In the Outer Hebrides, the process of creating a community economy "through and with" the wind is embryonic. A number of land owning communities have received planning permission for small-scale wind farms – among them Galson and Stòras Uibhist – and are in the process of building the turbines. For reasons already given, the NHT has now switched to pursuing the option of a small-scale hydro installation, together with the West Harris Crofting Trust (Chapter 5), but similar arguments may be made. Drawing on the experiences of these land owning communities, I want to argue in this part of the chapter that the furthering of a collectively directed economy – through wind or water – involves the repositioning of people both *vis à vis* the local community and *vis à vis* the networks that render the boundaries of that community porous.

For the NHT, which took a leadership role in pursuing the creation of a sustainable community economy through the sale of electricity to the national grid, local, community, control of the wind was seen as pivotal. The plan, until 2009, was to construct three turbines with a generating capacity of 2.5 MW, an amount that Scottish and Southern Energy had agreed to purchase for the national grid. The NHT could expect the turbines to provide an annual income, after loan repayments and running costs were deducted, of £250 000, an amount that could be invested in the community – in further renewable energy schemes, in insulating houses, in building new houses, in job creation, and in promoting small-scale businesses. This figure – of £100 000 per MW – contrasts with an offer from LWP of about £3000 per MW per annum for a rental payment and community benefit payment (MacDonald, 2010: 17). "In other words", writes former MP Calum MacDonald, "ten community owned turbines can deliver as much local benefit as three hundred corporate turbines" (MacDonald, 2010: 17).

For David Cameron, Chair, NHTC, speaking at the Community Energy conference held in Tarbert, North Harris, 24–26 April 2006, generating energy from the wind was "the one game on the block that [would] put money back in to communities … and we can't lose sight of that". "We could have gone with something much larger and lost control", he said on another occasion,

> but here the community retains control. The big thing is that we can do something [with this sum of money] … and can say to the huge developers, "we don't want you here". Owning the land is critical here. … It's a one off opportunity for the NHT at the moment. If [we] can marry traditional community things that happen with quite a modern industry, in a superb environment – these three things – if [we] can marry these – [it's] a good job (interview, 20 June 2004).

On a much later occasion, when the NHTC had just received planning consent for the wind farm at Monan, Cameron commented that the generation of electricity from renewable sources was "not just a nice green idea. [The turbines] open the possibility of re-invigorating communities where development options are few or non-existent" (cited in *WHFP*, 22 February 2008: 2). For one of the NHT's employees, this was "one of the most morally justified community projects" (interview, NHT Development Manager, 21 June 2005).

The argument was reiterated by Huw Francis, Executive Director of Stòras Uibhist, in an interview (10 June 2010). Commenting on how both income and the number of jobs linked to the growth of a community economy had increased since Stòras Uibhist took over ownership of the land, Francis referred to the importance of recycling money locally. His task, as he saw it, was to raise the GDP of the island but it was also "to run the business as if it [were] a department of social responsibility". "The measure of success was not net profit, but net money in the local economy [through] delivering a social agenda", he said. With ownership of the land and their own financial resources, he continued, the community could determine their own priorities. The proposed wind farm at Loch Carnan, South Uist, consisting of three turbines with a total generating capacity of 6 MW expected to bring in an annual income stream of £600 000 per annum, would allow a number of ambitious projects. The scheme was fully supported by the community – there had been no objections – as well as by all the agencies whose agreement was required for planning permission (Francis, interview 10 June 2010). One particularly vital undertaking, made all the more urgent by the rise in sea level occasioned with climate change, and made more feasible through community land ownership of the entire estate (i.e. not just that part of the estate where the wind farm is to be located), concerns the maintenance of the drainage system down the western side of the island, much of which land is susceptible to flooding. Technically, responsibility lies with the Scottish Government's Environment and Rural Affairs Department, but the system has not been well maintained and has reached the point where flood gates need to be rebuilt. Other priority projects which could be funded through the wind farm include housing, much of which is below standard, and providing seed capital for small businesses, possibly along the lines of the Grameen Bank (interview, 10 June 2010).

In the cases cited, community consultations have been integral to the decision making process and thus the reconfiguration of a collective identity. In North Harris, as I have already indicated, these were accompanied by a ballot which indicated that 93 per cent of those voting were in favour of the wind farm (interview, NHT Development Manager, 5 December 2006). Directors have been explicit that, enthusiastic though they might be about the venture, "they won't push it against the community's wishes" (interview, NHT Director, 6 July 2004). "We don't want to be in the business of ramming something down the community's throat", said one director (interview, 6 July 2004). The membership has been kept informed of progress through the local newsletter, *Dè Tha Dol?*, through public meetings and through the annual general meeting. Now that the hydro scheme at Bun Abhainn Eadarra has replaced the wind farm at Monan in the NHT's aspirations, it is this scheme in which the community is now participating.

There have been minor perturbations, but these have been few and far between. One example concerns Galson, where, in the summer of 2009, a small number of objectors to the wind farm circulated a petition claiming that the proposed site was sacred ground. The land at Druim nan Carnan, it was asserted, was a burial ground for those who died in a battle between the Morrisons of Ness and the MacAulays of Uig in 1654 (MacInnes, 2009a; 1). The response of the Galson Trust's chair, Agnes Rennie, was that, as for many other sites in the area, this site had cultural value and as such had been discussed with archaeologists prior to the application for planning consent (MacInnes, 2009a; 1). As required in any such planning consent, an archaeologist would be to hand when the land was cleared (MacInnes, 2009a; 1). Both Rennie and a local resident, Alasdair Smith, who had worked the peats in the Druim nan Carnan area for 40 years, insisted that peat cutting had occurred on the site for generations and nothing had been found (MacInnes, 2009a; 1; Smith, 2009: 4). In Smith's view, those who authored the petition were seeking "to romanticise ancient tradition for their own ends" (Smith 2009: 4).

In the Outer Hebrides, as well as elsewhere in the Highlands and Islands, the process of people's repositioning with respect to the economy – the becoming of a collective subjectivity – also concerns renegotiating how the community binds itself to the extra-local. The Highlands and Islands Community Energy Company (HICEC) and now Community Energy Scotland (CES) have been key here. Established by Highlands and Islands Enterprise in December 2004 as a non-profit company limited by guarantee, HICEC was charged with the dual task of delivering the government's *Scottish Community and Householder Renewables Initiative* grant programme and in helping communities develop their own renewable energy projects with a view to securing revenue with which to build more sustainable futures (HICEC, 2006).[37] Touted as one of the success stories in the Highlands and Islands, HICEC provided financial and technical assistance to hundreds of community groups, for example to carry out feasibility studies and assemble funding packages. While this has been carried out on a case by case basis, HICEC placed great emphasis on building on the network of groups that existed in the area.

The organization was launched in Knoydart, in community ownership since 1997, and subsequently held workshops/conferences there and with other community groups – Gigha, Harris, Sleat (Skye) – in order both to promote informed discussion of different aspects of community energy and to foster inter-community networking so that community groups could learn from each other. This twin objective was evident, as an example, in the event hosted by the NHT, 25–26 April 2006, entitled *Community Energy: Leading from the Edge*. The conference brought together delegates from community groups stretching from Shetland in the north to Islay in the south. For David Cameron, leading the NHT's initiative, reflecting back on the first such conference in Knoydart, it was the opportunity to talk to others about community energy that "was worth all the conferences in the world – or at least all the formal bits of conferences". People made contacts and stayed in contact – through e-mail or telephone (interview, 14 December 2005). In Knoydart, he continued, "you talked to people who were actually doing it – not with the theoreticians' (interview, 14 December 2005).

By HICEC and by communities themselves, networking was seen as integral to the process of building sustainable futures. Among the early pioneers of community-generated electricity, North Harris, Westray (Orkney), Tiree and, on the mainland, Melness, worked together with HICEC to overcome teething problems. All were at a similar stage – in a "black hole" regarding how they would finance the turbines. Their cooperation, together with that of HICEC, who employed someone with financial expertise to work with them, proved to be very successful (interview, David Cameron, 14 December 2005). More recently, HICEC brought together the seven groups in the Outer Hebrides that were in the process of developing their own wind farms, not all of them land owning: Galson Energy, Tolsta Community Development and Horsader Community Development from Lewis, the NHT, and three groups from the Uists – North Uist Partnership, Stòras Uibhist and Comhearsnachd Bharraidh agus Bhatarsaigh. Experience as well as equipment (an anemometer mast) was shared, and community representatives met officials from the Comhairle, SNH and consultants with expertise on landscape and visual assessment and wind energy (MacInnes, 2008a: 7). Commented Carola Bell from Galson,

> We have met with the other Lewis and Harris groups before but getting all the Western Isles groups together is a first and we have gained strength in working together and sharing experiences. By learning how other groups have overcome difficulties [accessing the grid, planning requirements ...], this will also save our group unnecessary work in the future. There's some innovative stuff going on out there in communities in the Western Isles (cited in MacInnes, 2008a: 7).

Speaking in early 2008, when the NHT had finally received planning permission for the proposed wind farm at Monan, Alistair Macleod, Development Manager for the NHT, was enthusiastic in his praise of HICEC:

> As a community, it would have been almost impossible for us to get to this stage without HICEC. What they have done over the last three or four years is to give us a lot of financial help for the risk part of the work – the

various landscaping, ornithological and civil engineering studies we had
to carry out. They also gave us a huge amount of officer help (cited by
Russell, 2008: 11).

Communities were not alone in having to find their way through the
labyrinthine requirements of the planning process or in facing manufacturers, uncertain of what the market in turbines was (Russell, 2008: 11).
HICEC's role in taking much of the risk out of the process of developing
a proposal for a community owned wind farm has been critical.

In an energy context of rising costs and a growing emphasis on carbon
mitigation on the one hand, and the growing demand for the work
HICEC was doing on the other, those responsible for the formation of
HICEC argued the case for extending the remit geographically to include
the whole of Scotland. A new company, Community Energy Scotland,
with a Scotland-wide remit, independent of HIE, became operational in
August 2008. CES has both charitable status and, since the summer of
2010, its own trading subsidiary, Community Energy Scotland (Trading),
which will allow it, eventually, to become financially self-reliant through
its own generation of energy from renewable sources. Referring to the
new organization, Executive Director Nicholas Gubbins emphasized
CES's position of seeing "community groups in the forefront of measures
to promote carbon mitigation as well as helping people to adapt to lower
energy lifestyles" (cited by Russell, 2008: 11). For that reason, he said, CES
had "a new purpose": "to build confidence, resilience and wealth at community level in Scotland through sustainable energy development" (Russell,
2008: 11; also, see CES, 2010: 1). Of fundamental significance, unlike
HICEC, CES is itself a community-based organization. Community members have voting privileges and can thereby influence the direction of the
new organization. They are central to its governance. The membership fee is
£10. To Alan Hobbett, chair of the new company, CES is not just "building
a new organisation, but a new movement"(CES, 2009: 3). Independent of
HIE, it can take a more overtly political role than had previously been the
case. With a growing membership base, it can more effectively engage in
lobbying to deal with issues that many communities face – problems with
grid connections being among the most contentious – as well as accessing
sources of funding from which HICEC was excluded. For the same reason,
it can command a stronger voice in such government consultations as that
on Renewable Electricity Financial Incentives, arguing the case for a retention of the Feed-in Tariff incentive (CES, 2010: 11).[38]

As a network, CES has been strengthened through such initiatives as
the bulk procurement and installation of smaller turbines, thereby

achieving economies of scale for its members (CES, 2009: 3). The significance of this facility becomes evident when it is recognized that the cost of turbines has increased greatly in recent years and that manufacturers find it more profitable to sell to large-scale producers than to small-scale enterprises in "remote" locations (*WHFP*, 4 May 2007: 7). Community Powerdown, funded through the Scottish Government's Climate Challenge Fund and working with 27 communities since 2008 to reduce dependency on fossil fuels, supports further the broadening of the membership base (CES, 2008: 4).[39] Galson is one of the sites where the CES is funding a staff position under Powerdown.

While I have focused thus far on the significance of a more community-centric economy through the generation of energy from the wind for the opening of new political possibilities, recognizing that this process has barely begun in the Outer Hebrides, recent research on Gigha, where the process has been underway for a considerably longer period of time, allows me to extend the argument. On the basis of a comparative study into attitudes towards wind farms, large-scale, corporate-led developments in Kintyre (on the mainland) and the three turbine community-owned wind farm on adjacent Gigha, Charles Warren and Malcolm McFadyen (2010: 209) show that the material advantage of community ownership of a wind farm cannot be divorced from "the positive psychological effect of community ownership" (Warren and McFadyen, 2010: 209). Whereas the residents of Kintyre showed an "apparent disconnection and (for some) alienation" towards the commercial wind farms, those in Gigha "revealed a strong sense of pride in and connection with 'their' windfarm project" (Warren and McFadyen, 2010: 209).

Warren and McFadyen (2010: 209) suggest that "the sense of ownership" experienced by the people of Gigha is "evocatively revealed" in the names given to the turbines. "The Three Dancing Ladies", they write, evoke ideas of gracefulness, of a pleasing aesthetics, of "a sense of belonging" (Warren and McFadyen, 2010: 209). The sense of ownership is reinforced by the Gaelic names assigned to the turbines: *Creideas*, *Dòchas* and *Carthannas* (Faith, Hope and Charity), which words contrast so vividly with the language of "rape" or "desecration" of opponents of wind farms on the mainland (Warren and McFadyen, 2010: 209). This naming of the turbines, Warren and McFadyen (2010: 209) continue, "implies that they are perceived as a physical embodiment of community cohesion and confidence". Moreover, while such naming may be "novel" in present-day Scotland, the practice has been the norm in the Netherlands and used to be "commonplace" in the early years of wind farms in Sweden, when each turbine was also assigned a gender (Warren and McFadyen, 2010: 209).

The naming of a turbine may be unprecedented in Scotland, but there is a lengthy history of naming identifiers of place within the landscape (Chapter 2). It is through the density of place-names that evidence of people's intricate relationship with the land is revealed. The names are a means through which people negotiate socio-natural boundaries. By naming the turbines on Gigha, people re-claim past practice, identifying features of the land with which they are closely associated and which, through being assigned a name, become one of the ways they attach themselves to place, now at the beckoning of a new collective idiom. As such, the turbines are implicated in the production of a "differential geography" (the phrase is Castree's, 2004: 136), in which people reposition themselves *vis à vis* property and nature in such a way as to allow a land owning community to "make" its own place (Castree, 2004: 136). The repositioning of people – or process of resubjectivation, to recall Foucault's (1990: 28) term – is prompted at the point where the norms of property are disrupted and the political possibilities of a new more social nature open up. The interruption proceeds through a refusal on the part of communities to separate the politics of property – who owns the land – from the politics of nature – who can capture the wind. Commodification proceeds but, allied to a community ethic centred on the process of a commoning of the land, the wind's enclosure provides the means of working towards "a politics of the otherwise" (Gibson-Graham, 2003: 53) and an "economy of generosity" (Gibson-Graham, 2006, 150).

It may be appropriate, in conclusion, to recall that, while community owned wind farms may be aligned with "neo-communitarian discourses of local participation and empowerment" (Warren and McFadyen, 2010: 205, citing the work of Walker and Cass, 2007, and Walker *et al.*, 2007) and an interest on the part of government in making "community renewables" (the term is from Walker and Devine-Wright, 2008, cited by Warren and McFadyen, 2010: 205) an integral part of government policy, for land owning community trusts their meaning is more radical, tied as capturing the wind is to the realization of an agenda of community land ownership. It is not simply a matter of "energy citizenship" (Devine-Wright, 2007, cited by Warren and McFadyen, 2010: 205) – signifying "new, active and participatory connections with energy generation and supply" (Warren and McFadyen, 2010: 205) rather than connections with the grid as "end-of-wire captive consumers" (Walker and Cass, 2007: 464) – important as this move undoubtedly is. It is about a very different "landscape of power" (Pasqualetti *et al.*, 2002: 3), where claims to a new energy commons are allied to the working of a new property commons.

Notes

1 Whereas the average "community benefit" from corporately owned wind farms in the UK amounts to about £4000 per installed megawatt of generating capacity – which would bring in about £2500 per annum if Gigha's turbines were owned by a corporation – Gigha's three turbines bring in 40 to 50 times as much (Hunter, 2012: 132).

2 Eigg, for example, had been assisted by Community Energy Scotland through CARES, the Scottish Government's *Community and Renewable Energy Scheme*. As discussed later in the chapter, CES grew out of Highlands and Islands Enterprise's Highlands and Islands Community Energy Company (HICEC). CES became operational in 2008. It has also had responsibility for running HIE's *Communities Renewable Energy Support Programme* (CRESP) and for developing community-level income generating projects from renewable sources. It supports communities' applications to the Big Lottery's *Growing Community Assets* fund.

3 On the basis of calculations of wind speed by Troen and Petersen (1989), cited by Warren and Birnie (2009: 104), Scotland has one-quarter of the onshore and offshore wind resources of Europe, the highest of any European country. The strategic importance of the Outer Hebrides in this respect lies both in the strength of the wind (and wave and tidal regimes) and in their consistency (Aquaterra, no date).

4 There is, at present, an awkward situation in the UK, where energy is classed as a "reserved" matter under the terms of devolution, whereas the promotion of renewable energy and the protection of the environment are devolved to the Scottish Parliament. In practice, this means that the Westminster Parliament is responsible for regulating the generation, transmission, distribution and supply of electricity as well as matters to do with non-renewable sources of energy (coal, oil, gas and nuclear). Holyrood has jurisdiction over the planning process, including issues that pertain to the environment, licenses for the generation and transmission of electricity, the conservation of energy and measures to mitigate the effects of climate change. In 2008, the Scottish Government revised its 2003 target of 18 per cent of electricity to be generated from renewable sources by 2010 and 40 per cent by 2020 upwards, as its target for 2010 had been met by 2007 (Scottish Executive, 2008a: 2). The new targets are to generate 31 per cent of electricity from renewable sources by 2011 and 50 per cent by 2020 (Scottish Government, 2008b). In contrast, the UK aimed to produce 10 per cent of its electrical energy from renewable sources by 2010 and 20 per cent by 2020 (CnES, 2005a: 323). The UK target is in line with that indicated in the EU Renewables Directive (Directive 2001/77/EC, as cited by CnES, 2007a: 120). As a signatory to the Kyoto Protocol in 1997, the UK made a commitment to reduce CO_2 emissions by 20 per cent by 2012 (CnES, 2007a: 119). In the Climate Change (Scotland) Act 2009, the

Scottish Government gave a target of decreasing greenhouse gas emissions by 80 per cent by 2050.

5 This very centralized model for developing renewable energy contrasts with that of the "community energy model", which has been widespread in other European countries, particularly Denmark, Germany, the Netherlands and Sweden (Warren and McFadyen, 2010: 206). Here, many renewable energy schemes are both "funded and controlled by farmers and local wind cooperatives" (Warren and McFadyen, 2010: 206), although, as Warren and McFadyen (2010: 211) go on to note, in Denmark at least, this situation is changing (also see Meyer, 2007). In the UK, there has been a recent change in policy whereby "community renewables" are now part of "mainstream energy policy" (Walker and Devine-Wright, 2008). Having noted this trend, however, Gordon Walker and Patrick Devine-Wright caution that the meaning of "community" in so far as energy is concerned is far from uniform. They argue that, the greater a community's involvement in a project, the greater the degree of acceptance and support, not just for a specific project, but also for renewable energy more broadly (2008: 499). They conclude their study by pointing out that, in contrast to practices in England, Scotland has demonstrated "a more concerted commitment to the deeper meaning of community in 'community renewables'" through the provision of advice and practical support at the local level through the Scottish Community and Household Renewables Initiative (Walker and Devine-Wright, 2008: 499; also see Walker et al., 2007: 71).

6 In October 2008, the Swedish, state-owned company, Vattenfall took over AMEC Wind Energy. This takeover gave the Swedish company control of all AMEC's wind projects except for the proposed wind farm on Lewis, a joint venture with British Energy, which, in any event, by that time, had been refused by the Scottish Government (WHFP, 17 October 2008: 9).

7 Peat slides had become a significant environmental concern following the European Commission's action against the Government of Ireland when a major peat slide took place at Derrybrien, County Galway, in October 2003, an event caused by work on a large-scale wind farm project (CnES, 2005a: 176). This "bogalanche" occurred on a 71 turbine wind farm on the summit of Cashlaundrumlahan in the Slieve Auchty mountains (Phillips, 2005). About $450\,000\,\mathrm{m}^3$ of the peat "flowed" down the mountain, causing a major environmental disaster (Phillips, 2005).

8 Considerable controversy surrounds the claim that a wind farm on the scale proposed by LWP is necessary in order to increase the export of electricity from the islands through an interconnector. See TNEI Services (2007) for an assessment of the options for grid connections for Orkney, Shetland and the Outer Hebrides, a study commissioned by Highlands and Islands Enterprise. A report by N. C. Scott of Xero Energy (2007) puts the study by TNEI in the broader UK context.

9   LWP (2004: 21) also claimed significant carbon savings from the development. With the revision downwards in number of turbines in the Addendum to the Environmental Statement, LWP's contribution to meeting Scotland's renewable energy target of 6 GW in 2020 would drop to 11 per cent (CnES, 2007a: 121).

10 It is interesting to note that Banks also refers to Paragraph 44 of NPPG 14, albeit not, in this case, as the determining paragraph. The Habitats Directive, this paragraph states,

> does not impose a general prohibition on development in or adjacent to Natura 2000 areas. Many wildlife species and habitats readily co-exist with human activity, and they may well rely on it. Thus, for the most part, uses which have continued sustainably over many years, and may have contributed to the high conservation value for which the area is recognised, will accord with the aims of the Directive and may continue unchanged (cited in CnES, 2005a: 81–82).

11 The Western Isles Development Trust was set up in 2004 with a broad remit to support, through various means, "all types of development that will be of economic, educational, environmental, cultural, social or recreational benefit to the people of the Western Isles" (Western Isles Development Trust, 2005: 1). A more particular objective is to set up a fund with a view to achieving such development by distributing "the impacts and benefits of renewable energy developments" across the Western Isles (Western Isles Development Trust, 2005: 1).

12 The Addendum to the Environmental Statement of 2005 indicates annual payments of £3.4 m in rent and £1.85 m in community benefits. It also states that communities may opt to take a "community equity stake" of 15 per cent in the wind farm instead of community benefit payments. The community would have the additional option of investing a further 5 per cent in the wind farm (CnES, 2007a: 96). Community benefit, as well as such matters as the provenance of goods and services (in order to ensure that the Western Isles receive the maximum benefit from the wind farm), are spelled out (in draft form) in what is called a Section 75 Planning Agreement in the Addendum to the Environmental Statement of December 2006, to be signed by the Comhairle, LWP and the land owners (CnES, 2007a: 125–154).

13 Iain MacIver, factor of the Stornoway Trust, made this trust's continued support for the wind farm clear in a letter to Jim Mather, Enterprise and Energy Minister, in October 2007. Cited in the *WHFP*, MacIver writes that "the trust's pro-wind farm stance appears to enjoy the endorsement of most of its 11,000-strong electorate with their emphatic support in 2004, and again this year, when pledged supporters of the wind farm were re-elected to serve as trustees" (*WHFP*, 5 October 2007: 3).

14   It is worth noting that, at the time of their response to Lewis Wind Power, SNH were co-funding with the Royal Society for the Protection of Birds research that produced a "Bird Sensitivity Map to provide locational guidance for onshore windfarms in Scotland" (Bright *et al.*, 2006, 2008).

15   LWP's Environmental Statement, using data from Macpherson Research (2003), indicated that the tourism sector was worth £39.3 million to the Outer Hebrides, "a significant proportion of the local economy" (cited in CnES, 2005a: 286). LWP estimated that the proposed wind farm "would have a significant but small-scale direct negative effect through discouraging some tourists from visiting Lewis", particularly those who were drawn "by the seclusion, wildness and natural habitats" of the island (cited in CnES, 2005a: 289, 290). In contrast, the number of people visiting for the purpose of business "could actually grow" (cited in CnES, 2005a: 290). For further assessment of the impact of wind farms on tourism in Scotland, see Riddington *et al.* (2008).

16   Of the other community councils where a ballot was taken about LWP's proposed wind farm prior to the CnES meetings of June 2005, 67.55 per cent voted against in Barvas and Brue, 73 per cent in Shawbost, 55 per cent in Sandwick and 67 per cent in Laxdale (CnES, 2005a: 579–581).

17   The meeting ended with a vote that endorsed the presentation and with a vote of confidence in the NCC to take the community's decision forward. In effect, as an article in the *WHFP* (24 June 2005: 5) alleges, the community council was "forced into a U-turn". The move resulted in the resignation of the chair and three other members of the council (*Stornoway Gazette*, 30 June 2005: 4).

18   For RSPB's rebuttal, see George Campbell's letter, as Director, RSPB North Scotland, to the Editor, *WHFP* 2 February 2007: 11.

19   In October 2007, BMP submitted another, smaller, wind farm, Feiriosbhal, with 16 turbines and a generating capacity of 48 MW to the Comhairle for planning permission, in the event that the proposal Muatheabhal Wind Farm was refused by the government (BMP, 2007: 1). With a generating capacity below 50 MW, as determined by the Electricity Act, Section 36, a wind farm of this size fell under the Comhairle's jurisdiction. The proposed turbines occupy the same locations as the 16 most easterly turbines in the larger proposal (BMP, 2007: 1). In the event, the Comhairle approved a 13 turbine, 39 MW, version of this smaller proposal in July 2008 (MacInnes, 2008b: 1). To one critic, the proposal indicated the "salami slicing" of the larger project, which was being cut into parts "in order to circumvent prior assessment of significant environmental effects prohibited by the 1999 regulations [i.e. The EIA (Scotland) Regulations 1999]" (e-mail from Jeremy Carter to Munro Gold, 29 November 2007, in CnES, 2008a). While SNH lodged no objections (letter from Mark Macdonald, Area Officer, Stornoway, 21 February 2008, in CnES, 2008a), Helen McDade, Policy Officer for the John Muir Trust, objected on the grounds that the turbines lay within one of "the remotest areas of wild land in Scotland", one of four areas with a distance of

over 8 km from a road (in CnES, 2008a). With the announcement that the government approved the larger wind farm, this proposal was eclipsed.

20 Development rights were, in other words, separated from the land and registered in a legally separate company. Were the land to be sold, it would be sold without these rights and thus a community would be unable to exercise the rights it anticipated with ownership.

21 The Graham family had owned the estate since 1924, when the Isle of Lewis was divided into several estates following Lord Leverhulme's departure. Galson Estate Limited, the private land owning body, had four directors at the time of the negotiations – Sandy Graham, Ann Graham, Jean MacMillan and Alasdair MacRae, all of whom resided on the island (*WHFP*, 18 February 2005: 3).

22 One member was of the view that the steering group's hands were tied in that, were it to express a view against the turbines, its ability to access external funds would be compromised as it would not be seen as favouring economic development (interview, member of the steering group, 15 December 2008). Evidence of the tension comes from the elections held when the steering group was replaced by the community trust. Whereas it is commonly the case that the directors of the steering group are elected to the new community body, in this case 30 people stood for 10 positions (interview, member of the steering group, 15 December 2008; *WHFP*, 30 September 2005: 3). Many of those on the steering group were re-elected, but not all (interview, member of the steering group, 15 December 2008).

23 The purchase price for the land was £600 000 (*WHFP*, 22 December 2006: 11).

24 Prior to this arrangement – i.e. had the sale of the estate together with the development rights not proceeded – crofters with a share in the common grazings where the turbines were located would have received 40 per cent of the sum agreed with Lewis Wind Power, 40 per cent going to the landlord, and 20 per cent being allocated to the "wider community" through a trust (*WHFP*, 11 January 2003: 3).

25 A spokesperson for the Pàirc Trust explained that the effect of the interposed lease was to place an "intermediate landlord" between Pàirc Crofters Ltd and the crofting community (interview, 3 July 2008). Pàirc Renewables would be required to pay a rent of £1000 per year to lease the entire estate for a period of 75 years; the Pàirc Trust, were it to purchase the estate, would thus be entitled only to collect this rent (interview, 3 July 2008).

26 The Pàirc Liaison Group (also referred to as the Pàirc Windfarm Liaison Group) had been set up in 2003 with a remit "to conserve and regenerate the Pàirc area of Lewis for the benefit of the Pàirc community" (*Stornoway Gazette*, 8 January 2004: 3). Its focus was on assessing the benefits for the community both of the siting of the wind farm on the estate and of community purchase of the land (*WHFP*, 29 August 2003: 2). The Pàirc Trust replaced the Liaison Group at the end of 2003 in order to pursue negotiations about the wind farm and land purchase on the community's behalf (*Stornoway Gazette*, 19 February 2004: 3).

27   These calculations were based on SSE's proposed wind farm of 125 turbines
      (*WHFP*, 22 August 2003: 9). The Comhairle's planning officers recom-
      mended a reduction to 57 in 2007 on the grounds of concerns about the
      negative effects on the landscape, visual amenity and proximity to dwellings
      (*WHFP*, 30 November 2007: 9). In April 2009, members of the Comhairle's
      Environment and Protective Services Committee recommended that a 26
      turbine wind farm be approved, bringing the total generating capacity down
      from 205 MW with 57 turbines to 94 MW (*Stornoway Gazette*, 23 April
      2009: 18). The Scottish Government has responsibility for the final decision.
28   The *West Highland Free Press* (3 June 2005: 1) gave the population of
      Pàirc as 400 in 2005.
29   Despite having secured a major award from the Big Lottery Fund
      (£900 000), the NHT decided to drop the plans for a wind farm at Monan.
      Turbulence at the site had always been a problem – caused by the effect on
      the wind of the mountainous terrain. The risk became more apparent as
      negotiations to procure turbines proceeded, leading David Cameron, Chair
      of the North Harris Trading Company, to state that, frustrating though
      such a decision might be, it was "based on good commercial reasoning"
      (cited in *WHFP*, 25 December 2009: 1). The trading company is now ex-
      ploring two small-scale hydro installations (Chapter 5).
30   As a wholly owned subsidiary of the NHT, the remit of the North Harris
      Trading Company is to progress all commercial activities of the trust. All
      income goes back to the trust. At present, the NHTC board comprises three
      NHT directors, three members of the community and one member from
      outwith the community.
31   These included the *Landscape and Visual Impact Assessment* by Horner
      and MacLennan (2006), an *Archaeological Walk-over Survey* (Knott and
      Harris Archaeology Group, 2006), an *Ornithological Survey and Assessment*
      (Haworth Conservation Ltd., 2006) and statements from Highlands and
      Islands Airports Ltd and Defence Estates.
32   In addition to potential sites for on shore wind farms, the study focused on
      the possibilities for the generation of electricity from offshore wind energy,
      wave power, tidal power and small-scale hydro schemes; to a lesser extent,
      the consultants considered solar energy, biomass and geothermal sources
      (West Coast Energy, 2004).
33   As required by The Town and Country Planning (Notification of Applications)
      (Scotland) Directive 1996, in a situation where a statutory authority raises
      an objection and a local council is "minded to grant planning consent", the
      Comhairle had to forward the case to the Scottish Government (CnES,
      2007b: 130).
34   For an assessment of the economic impact of the proposed wind farm by
      the Comhairle's Economic Development Department, see CnES, 2007b:
      112. This report concludes by saying that a "significant impant" is expected
      at the local level,

where both the economy and community face significant challenges. The
project will place new resources in community hands; will empower the
community to move forward their own priorities and will open new opportunities for entrepreneurship and innovation (CnES, 2007b: 112).

35 While Cameron does not specify the threat to Gaelic occasioned by further
social and economic decline, the claim that Harris is of national importance
could be additionally strengthened by identifying the centrality of Harris to
the Gaìdhealtachd more generally and to the language more specifically.

36 This reconfiguration of nature, producing the wind as a commodity but
tied to a community ethic, was supported by the John Muir Trust. The JMT
had backed the NHT's bid to own the land in 2003, providing £100 000
towards the purchase price of the estate and, since that time, had held an
invited directorship on the NHT board. Mick Blunt, JMT officer for the
Outer Hebrides, made clear in a letter to the Comhairle in 2006 that his
organization was behind "the development of small-scale, sensitively sited
renewable energy schemes in areas adjacent to existing settlements, which
demonstrate that renewable energy may be sourced without significantly
impacting on wild land" (cited in CnES 2007b: 109). He emphasised the
significance of the project for the regeneration of the estate – with respect
to local housing, economic development and the environment (cited in
CnES 2007b: 110). This position was reiterated by JMT Chief Executive
Nigel Hawkins in a letter to the editor of the *WHFP* dated 29 February
2008. Small-scale, community-owned, wind farms, he proposed, provided
a "clear alternative" to such large-scale projects as that of LWP in a comprehensive energy strategy that sought both to decrease consumption and
to promote the production of electricity from renewable sources. In contrast to a large-scale wind farm, a small-scale community owned wind
farm, to paraphrase, would mean that "all the profits [would be] ploughed
back into local development" (in *WHFP*, 29 February 2008: 13).

37 HICEC itself grew out of HIE's Community Energy Unit, operational since
December 2002. In the words of Nicholas Gubbins, head of community
regeneration at HIE, the unit "has seen a quiet revolution taking place as
organizations grasp the opportunities to install and develop a range of
renewable energy technologies" (cited in *WHFP*, 21 January 2005: 5).

38 The UK Government introduced the Feed-in Tariff as a measure to
encourage the generation of electricity from renewable sources for small-
scale schemes under 5 MW. With a re-interpretation of the rules following
Westray and Tiree's successful application to the Big Lottery for funds to
finance the turbines, it is now no longer possible to access both public
funding and benefit from the Feed-in Tariff. It is this matter that CES is
contesting (interview, David Cameron, 30 March 2011). The Feed-in
Tariff exists alongside the Renewables Obligation, introduced by the UK
Government in 2002 as an incentive to large-scale suppliers to generate an

increasing percentage of their electrical output from renewable sources. Large-scale suppliers generating "green electricity" receive a Renewable Obligation Certificate (ROC) for each MWh of electricity they produce. ROCs have value for trading (CES, 2010: 11). This approach differs from the tariff-based approach of other member countries of the EU, which relies on a fixed price per megawatt generated (Warren and Birnie, 2009: 103).

39   Another instance of networking has been the Outer Hebrides Energy Co-operative or Eco-Heb as it is known. The membership comprises Galson Energy Ltd, Horshader Community Development Ltd., Tolsta Community Development Ltd., Point and Sandwick Power and the North Harris Trading Company Ltd. (*WHFP*, 21 August 2008: 3). By co-ordinating their members' efforts, Eco-Heb aims to achieve "the maximum local benefit in local jobs and purchases", the secretary, Calum MacDonald, has stated (*WHFP*, 21 August 2008: 3). Eco-Heb took an active role in protesting Scottish and Southern Energy's reversal of its decision to provide early grid connections for island communities (*WHFP*, 7 May 2010: 1, 2).

5

Working Places

Introduction

> Community [land] ownership offers endless opportunities for everybody involved. ... [And] the ideas are only constrained by our own imagination. ... The sense of community and the sense of place expressed in songs and poetry is not something that belongs to a past age. It is still very much alive in communities across the Highlands and Islands today. Community ownership of land is the ultimate realisation of the link between people and land, but it is not a museum piece to be admired in a spotlight. It is a living, breathing entity and as modern day owners it is our responsibility and our challenge to nurture the land ... for generations in the future. ... As my grandmother said in the shelter of the walls in Galson, "We were not coming on a temporary basis, we were here to stay" (Rennie 2010: 14–15).

With these words, Agnes Rennie, Chair of Urras Oighreachd Ghabsainn/ Galson Trust, concluded her delivery of the Seventh Annual Angus Macleod Lecture on the stormy evening of 28 October 2010 in Pàirc School, Gravir, South Lewis. Honouring a man passionate about the land and the need for land reform, Rennie identified some of the possibilities opened up by community ownership. These ranged from new initiatives in providing housing, carbon reduction and encouraging new businesses to marketing croft produce. She emphasized the possibilities created when community trusts worked with each other – renewable energy

Places of Possibility: Property, Nature and Community Land Ownership, First Edition. A. Fiona D. Mackenzie.
© 2013 A. Fiona D. Mackenzie. Published 2013 by Blackwell Publishing Ltd.

initiatives on land or offshore and the creation of walking trails through the islands and the mainland. To the list of possibilities could be added Stòras Uibhist's ambitious £9.5 million project to redevelop Lochboisdale Port of Entry, funding for which is now in place. The redevelopment is expected to create 126 jobs, construct two causeways, provide land for businesses, upgrade the pier and facilities to support fishing and encourage yachting, and allocate land for public and private housing (*WHFP*, 7 January 2011:1, 20 March 2011: 1). The project had been identified as key to reversing the economic and social fortunes of the South Uist Estate at the time of the buyout in 2006.

My focus in this chapter is on a series of initiatives carried out by the North Harris Trust – much smaller in scale than those to which Stòras Uibhist aspires – which demonstrate further the political possibilities that are opened up by community land ownership. In each case, I show how the interruption of a process of enclosure and privatization creates a new way of seeing the estate and reconnecting people to place. I examine how, through measures to curb the growing speculation in the sale of croft land for house sites, efforts to reduce the community's carbon footprint, the encouragement of local food production, the conduct of archaeological research and the restoration of a network of footpaths, the North Harris Trust produces a counter-visuality to that of an estate previously constituted through the norms of private property and a nature from which people were separate. These initiatives challenge the "givenness" of the land. They suggest a working of power that, by producing a commoning of the land, opens up the land to new performances and people's repositioning *vis à vis* place.

Accommodating places

Their white walls and bright red doors and window frames a vivid contrast against the bare rock and rough heather moorland beyond, the eight houses at Ceann An Ora mark the culmination of years of effort on the part of the North Harris Trust to support the building of affordable housing on the estate (Photograph 5.1). The houses – a mix of one, two and three bedroom homes – have been built through Hebridean Housing Partnership and Tighean Innse Galle and were available for rent in early 2011.[1] Next to this site, the NHT has made available for sale three fully serviced, private, plots. It has also identified six further sites for sale, three in Tairbeart, two at Cliasmol and one at Bedersaig. With the exception of the two sites at Cliasmol,

Photograph 5.1   New housing at Ceann An Ora.
Source: The North Harris Trust. Used by kind permission.

which are on NHT land outwith crofting tenure, the plots are located on land under crofting tenure. Here, the NHT has had to negotiate for the resumption of the land with members of the respective crofting townships. Income from the sale of the land under crofting tenure will be shared equally between the trust and the crofters. The trust itself also owns property which it rents out. The two quayside cottages at Abhainn Suidhe – near the castle and previously belonging to the private owner of the estate – now provide affordable rental accommodation, and the NHT is looking into the possibility of converting the adjacent peat store into a small rental flat. Two further, two-bedroom, flats, located on the floor above the new NHT office building in Tairbeart, were made available for rent in mid-2011.[2] Visually, the new buildings and the rehabilitated cottages call attention to property a-working with community ownership of the land. As I show below, they are integral to the constitution of a counter-visuality that signals a more socially just working of the land than previously prevailed.

In North Harris, as elsewhere among community land owning trusts, housing, and specifically affordable housing, is seen as an integral component of a strategy to reverse the tide of outmigration of, particularly, youth and to build a more socially, culturally and economically sustainable future (Logie, 2007: 16). It has been a priority for the NHT since its inception. House prices have escalated in recent years as more and more houses have been converted into holiday or second homes and incoming retirees have offered prices that are out of the reach of young, local, people, a reality that is all too evident throughout the Highlands and Islands (Mackinnon, 2005: 12; Jedrej and Nuttall, 1996; Logie, 2007).[3,4] It is also the case that the available housing stock has been depleted by the conversion of houses into self-catering units in response to the growing tourism market.[5] Sean O'Drisceoil's (2009) *Lewis and Harris Rural Community Housing Pilot* report detailed the extent of the problem of lack of affordable housing in Harris as well as in the districts of Uig, Bernera and South Lochs in Lewis, sites chosen on the grounds that they have a large percentage of holiday homes and house prices beyond those affordable by young families. While identifying the combination of factors that has led to the current situation, the report notes the disturbing fact that no affordable housing had been built in Lewis in the twenty-first century outwith Stornoway. Nor had any such housing been built on Harris during that time period.[6]

On the North Harris Estate, where, with the addition of the Loch Seaforth Estate in 2006, 46.9 per cent of the land is under crofting tenure, and more generally in the Outer Hebrides, where 77 per cent of the land is so designated (Logie, 2007: 9) and where croft land is generally more accessible to existing infrastructure than non-croft land, the NHT recognizes that the availability of croft land is critical to achieving and sustaining affordable housing objectives (NHT, 2009b: 1). This has been evident, for example, in the NHT's negotiations regarding the resumption of common grazings for the houses at Ceann An Ora. On a broader front, Derek Logie's (2007) study, *Houses on Crofting Land*, carried out for the (then) Scottish Crofting Foundation, makes clear that it is the supply of land that is central in increasing and sustaining affordable housing and that the provision of croft land is critical in this respect. He cites from the *WHFP* (July 2005) to make the point:

> social housing in West Highland communities – formerly provided by
> local authorities – has hitherto been ancillary to that other "affordable"
> option, housing associated with croft tenancies, which has acted as a
> highly-effective bulwark against a market controlled solely by money. ...

[W]hen anyone looks for an answer to the "affordable" housing shortage in crofting areas, they overlook the fact that crofting tenure – if properly administered and regulated – provides precisely that solution (cited by Logie, 2007: 43).

One example from Lewis demonstrates how this may be done. In this case, the Tong and Aird Grazings Committee with the cooperation of the community land owning Stornoway Trust has implemented a plan to provide affordable housing for young, local, people by identifying potential sites on less agriculturally valuable common grazings (*WHFP*, 21 October 2005: 1). Each site is sold at a fixed price which is considerably less than the market price. The income, once legal and administrative fees are deducted, is then shared equally between the grazings committee and the Stornoway Trust (*WHFP*, 21 October 2005: 1).The grazings committee then invests its share in the crofting township, for example through the purchase of agricultural equipment, and in initiatives of value to non-shareholders such as making a donation to the village hall (*WHFP*, 21 October 2005: 1, 3). The scheme led to the building of ten houses in the three years up to 2005, with three more houses under construction in 2005 (*WHFP*, 21 October 2005: 1). Of the two difficulties encountered, one concerned infrastructure – distance to a sewer in the case of more recent building, a cost shared by the buyer and seller; the second concerned eligibility (*WHFP*, 21 October 2005: 3). Interviewed by the *WHFP* (21 October 2005: 3), Angus MacLeod, grazings clerk for Tong and Aird, spoke about the difficulty of defining what counted as "local" in prioritizing applications for housing sites. "Does the fact that one of their grandparents came from the area qualify a person as local?", he asked (*WHFP*, 21 October 2005: 3).[7]

Such was the success of the scheme that an editorial in the *WHFP* (21 October 2005: 17) called on government ministers to recognize that such initiatives as Tong's were

totally compatible with the crofting system and largely incompatible with the proposed free market [in the new (2007) crofting Bill] and then [to] abandon their alarming rhetoric about the need to get rid of a "protectionist mentality" surrounding crofting tenure.

The editorial continues:

The Bill, by licensing individuals to buy and sell tenancies without regard to the crofting interest, would fatally undermine the whole system, including the intelligent thinking that is now taking place on how to address the housing shortage (*WHFP*, 21 October 2005: 17).

However, elsewhere, it is precisely this land, under crofting tenure, that has become the subject of heated controversy insofar as providing sites for housing is concerned. A burgeoning market in croft land, facilitated by the Crofting Reform Act 1976, which gave crofters the individual right to decroft their land (Chapter 2), has pushed the price of plots of land well beyond the reach of young people or those with limited incomes.[8] Croft land, and generally the more agriculturally productive inbye land, has in effect been sold as sites for houses.[9] One particularly notorious case involved an absentee crofter's application to the Crofters Commission to decroft inbye land in Taynault, on the mainland, in 2005. As planning permission had already been granted by the local planning authority, the Commission stated that it had no option but to approve the application (Logie, 2007: 4). Opposition to the decision was all the more strident as the decrofted land was to be the site of executive style houses, valued at £365 000 for a four bedroom house, rather than affordable housing (Logie, 2007: 4). An editorial in the *West Highland Free Press*, dated 18 March 2005, made the case succinctly:

> There is limited point in housing associations building new developments on the fringes of Stornoway, Portree or Balivanich if the demand for them is being exacerbated by the fact that crofts are being bought and sold to the highest bidders like pieces of freehold land. Either the legal framework of crofting tenure should be enforced or the pretence abandoned. ... [A]s an instrument for allowing people to stay in their own communities, crofting tenure is the bird in the hand and it is time to stop it flying away (*WHFP*, 18 March 2005: 11).

That crofting, which had in the past secured rural communities, was now the means through which those same communities were in the process of being eviscerated was recognized by the Committee of Inquiry on Crofting, chaired by Mark Shucksmith, whose final report was published in 2008. This committee had been established in response to the outcry against the Crofting Bill, introduced by the then Scottish Executive in 2006, which was seen as endorsing the growing free market in croft tenancies. The inquiry aimed to produce "a vision for the future of crofting" based on an assessment of the extent to which crofting contributes to "sustaining and enhancing the population; improving economic vitality; safeguarding landscape and biodiversity; and sustaining cultural diversity" (Committee of Inquiry on Crofting, 2008: 3). For these contributions to be maximized, the report asserts that "an appropriate balance between individual and wider interests must be struck" (Committee of Inquiry on Crofting, 2008: 19). "Ultimately", the report states, "unless wider interests, especially those of future generations, are given precedence over individual interests, crofting will disappear and the potential benefits of its contribution to sustainable rural development will be lost" (Committee of Inquiry on Crofting, 2008: 19).

Identifying as the most serious threat to the wider interest the burgeoning market in croft tenancies, the Committee of Inquiry on Crofting (2008) recommended a series of measures that would support the broader crofting interest rather than promoting individual gain. These included the more effective regulation of sales, assignations, and applications to decroft, residency requirements, and the working of the croft (Committee of Inquiry on Crofting, 2008: 8). To this end, the committee outlined changes in the structure of governance of crofting in order to set in train "greater local accountability and ownership" of the regulatory process (Committee of Inquiry on Crofting, 2008: 11). Paralleling these recommendations to curb the marketization of crofts and to support the sustainability of crofting for the future, and responding to the rapid rise in house prices, which placed houses out of financial reach for local and particularly young people, the committee detailed measures to reverse these trends. Centrally, and controversially, the committee stipulated that all croft houses would be tied to

occupancy (i.e. residency) through a real burden [a condition attached to a property right], which will be deemed to be included in the conveyancing when next assigned or purchased. This would run with the land in perpetuity (Committee of Inquiry, 2008: 64).

A similar burden would be applied to new houses, whether these were located on decrofted land or not, and this burden would be part of the local authority's planning consent.[10]

In its response to the committee's report, and in a political context of growing polarization of the issue, the Scottish Government (2008c: 15) stated that it would replace the suggested burden with an "occupancy condition", to be imposed where houses were built on decrofted land.[11] However, in the ensuing legislation, The Crofting Reform (Scotland) Act 2010, the Government replaced the idea of an occupancy condition with a series of measures designed to address the issue of speculation. The first concerns extending what is known as the "landlord's claw back" from five years as stipulated in The Crofters (Scotland) Act 1993 to ten years (Crofting Reform [Scotland] Act 2010, Part 4, Section 41).

What this means is that, if a crofter sells the croft to someone outwith their family within ten years of having bought it, the landlord is entitled to a further payment (see Edwards, 2010). Second, the legislation strengthens the ability of the Scottish Land Court to refuse applications to resume land (i.e. to remove land from crofting tenure) (Crofting Reform [Scotland] Act 2010, Part 4, Section 42) and that of the Crofters Commission (to be renamed the Crofting Commission) to reject applications to decroft, even where planning permission had been given (Crofting Reform [Scotland] Act 2010, Part 4, Section 43). The grounds on which resumption and decrofting may be refused make clear the significance of the broader interests of crofting and "community":

(a) the sustainability of –
    (i) crofting in the locality of the croft or such other area in which crofting is carried on as appears to the [Land Court or Commission] to be relevant;
    (ii) the crofting community in that locality or the communities in such an area;
    (iii) the landscape of that locality or such an area;
    (iv) the environment of that locality or such an area;
(b) the social and cultural benefits associated with crofting (Crofting Reform Act 2010, Part 4, Section 42 [Land Court] and Section 43 [Crofting Commission]).

While it is clearly too early to assess the impact of the 2010 Act on the marketization of crofts and speculation in the housing market, there are a number of measures that community land owners can introduce to mitigate the current situation. As Derek Logie (2007: 47) notes, who owns the land has a bearing on the availability of house sites, and community land ownership (or other kinds of "not-for-profit" land owners) could be important in facilitating the reduction of the price of house sites. The availability of land for housing depends on negotiations between the land owner and the grazings committee, the final sale price being divided equally between the land owner and the shareholders in the common grazings, as I indicated was the case with the housing sites at Ceann An Ora. Community land ownership does not automatically mean that there will be a congruence of interests between the two parties, of course. There might be a conflict of interest situation where individual crofters sought to maximize the value of the land while the community trust sought to ensure its affordability. Such has not, as yet, been the case in North Harris. Moreover, as identified below, crofters who are party to a sale of land with a community trust would have the assurance that the land would not be sold on at an inflated price.

In addition to facilitating the greater availability of sites for housing, which is seen as critical in stemming the upward spiral in land values, community land owning trusts can work with two sets of measures to "keep a lid" on land and house prices (interview, NHT Development Manager, 5 July 2004). The first concerns the Title Conditions (Scotland) Act 2003. Identified as one of the initial 19 Rural Housing Bodies under Section 43(5) of the Act, the NHT has the right to create "a real burden which comprises a right of pre-emption in favour of a rural housing body", which is applied to the title to each housing plot it sells (Title Conditions [Scotland] Act 2003, Section 41A[1]).[12] The right of pre-emption can detail how the price of the property subject to the buy back will be calculated, should the housing body decide to exercise this right (Young, 2004: 6). In practice, the right of pre-emption allows the trust to buy back the plot and any buildings on it at the time of the first and at all subsequent sales. The objectives of this burden and conditions of sale, as outlined in the NHT's Land Release Policy (NHT, 2009b: 2), "are intended to preserve the Community interest in land sold for housing; to ensure that houses are actually built and that speculation on land purchased does not occur" (NHT, 2009b: 2).[13] All titles, continues the policy, "will contain a clause imposing the continuing right of pre-emption" which "would have to be exercised within 42 days" (NHT, 2009b: 2). All sales of land for housing may also have as a condition of sale the stipulation that the building of a house has to be completed within five years. If this does not occur, the NHT would retain the right to buy back the plot at the original sale's price plus the District Valuer's valuation of any "development" that had occurred on the plot (NHT, 2009b: 2). Where the trust had sold a plot below market value, if a subsequent sale occurred within 10 years, the seller would have to repay the discount to the trust (and common grazings committee if the plot was on that land), with interest (NHT, 2009b: 2). In adopting the burden and conditions of sale, the NHT makes clear in its Land Release Policy that it is "governed by social and community benefit considerations and will use any income generated from sales to re-invest in North Harris" (NHT, 2009b: 2).

Second, following the practice of Rural Housing Bodies elsewhere, the NHT is in the process of formulating an allocations policy for accommodation, another instrument through which it can promote the sustainability of a small community.[14] As an example, the Knoydart Foundation's allocations policy for permanent accommodation includes awarding points on the basis of residency and length of residency, whether the applicant was educated in Knoydart, whether they had work in Knoydart, or had skills that could contribute "to the economic sustainability of the area", whether they had dependent children, whether they were "fleeing or threatened by domestic violence" or were "affected by a disability" (Knoydart Foundation, 2006).

Gigha provides a second example. After decades of no investment in housing and a situation where 75 per cent of the housing stock was classified by the council as "below tolerable repair", the Isle of Gigha Heritage Trust turned the situation around, as discussed in Chapter 4. It was "a huge success story", recalled Peter McDonald of Fyne Homes, a housing association working in partnership with the Gigha Heritage Trust and "committed to the community-centred provision of affordable housing" (speaking at the Community Housing Conference, Gigha, 3 November 2006). Under the partnership arrangement with Fyne, the Gigha Heritage Trust retained control over the allocation of the houses, giving priority to those who either lived on the island or who had a connection with it (David McDonald, Development Manager for the Gigha Heritage Trust, speaking at the Community Housing Conference, Gigha, 2 November 2006). However, the provision of housing has also been used to attract those without such connections but with particular skills to the island (Satsangi, 2007: 42). In this way, the skill base has been diversified and the population has grown (Satsangi, 2007: 42). The population, which stood at 98 at the time of the buyout, had increased to over 150 by 2011 (Hunter, 2012: 133).

And on Eigg, the Isle of Eigg Heritage Trust has developed what is known locally as "the Eigg Roll" system (Macleod, 2009: 36). As explained by Maxwell Macleod, the land owning Trust "lends" without cost a site to anyone who wishes to build a house, the agreement being that the value of the plot will be returned to the trust in the event of a sale, ensuring that the community maintains ownership of the land and benefits from any increase in value (Macleod, 2009: 36). The process works as follows: once the building is completed, both land and house are valued and, on the basis of the proportionate value of each, a shared equity agreement is signed. Thus, if the house value is set at £60000 and the plot at £40000, the shared equity is 60/40, a proportion that comes into play if the house is sold (Macleod, 2009: 36). For trust member John Booth, "The great thing is that we [the trust] won't be selling off the family silver" (cited by Macleod, 2009: 36).Those who remain on the island for over ten years are allowed to increase their equity to 85 per cent. The trust has identified 20 such sites and is lending out the sites at a rate of two per year.

These examples lend support to Alastair McIntosh's (2009: 17) observations about the potential for community owned crofting estates to provide leadership with respect to the provision of housing, a situation he contrasted with that in Ireland. Invited to attend a community meeting called to consider the housing shortage on Clare Island in 2009, McIntosh quotes the Chair, who made it clear that individual land ownership – the model currently in place – contributed directly to the housing shortage: "Since we kicked the British out we've all been proprietors. We're now the cause of our own troubles". A young farmer at the meeting offered to give a housing plot from his own land to the community to help resolve the shortage, "but only if we developed a structure to stop it from being sold on to the highest bidder" (cited by McIntosh, 2009: 17). The offer was matched by an "English incomer". In contrast to the situation in Ireland, where land reform legislated for private ownership among small-holders, McIntosh (2009: 17) writes that crofting community ownership provides that structure – it "allow[s] the collective holding of land alongside private ownership of the 'improvements' such as the house". But, he continues, "most astonishing of all is that these [ideas] can no longer be caricatured as yesterday's worn-out ideas from a dying way of life". Elinor Ostrom's Nobel prize in economics, he emphasizes, "shows how community groups can successfully manage overlapping rights to shared natural resources. Crofting is precisely such a system. ... Ireland's Ryanair-like ethos of privatization now lies tangled in negative equity and family bankruptcy. Community-held crofting tenure is now what's at the cutting edge of Nobel-class economics" (McIntosh, 2009: 17).

Community action with respect to housing has been substantially supported by events such as that at which Peter McDonald spoke, facilitated through the Skills Development Programme of Highland and Islands Enterprise's Community Land Unit. Paralleling initiatives aimed at facilitating networking among land owning community trusts and the exchange of expertise by HICEC (in Chapter 4 I cited the example of a workshop on renewable energy held in Tairbeart in April 2006), the Community Land Unit backed the Isle of Gigha Heritage Trust's hosting of a conference, *Community Housing – Building for the Future* on 2 and 3 November 2006. In addition to community representatives, who themselves had built up substantial experience, a number of legal and other "experts" were invited to the meeting. Workshops were held on such topics as Rural Housing Burdens, partnering with a housing association, financing community land initiatives and green technologies. The conference had obvious tangible practical benefits – the exchange of

knowledge about the options available to support community-led housing initiatives and the opportunity for communities to work with each
other, to gain from each other's experience in negotiating these different
options.

The designation of croft land or land outwith crofting tenure for new
house plots, the right of pre-emption and an allocations policy for accommodation allow a community land owning trust to effect a measure of
what Castree (2004: 160) calls "place control". Bound up with others'
efforts to reverse historical and contemporary processes of dispossession,
the moves provide a means through which people can work towards the
reduction of inequalities with respect to the housing market. This is a
case of "situational pragmatism" (Castree, 2004: 163). Together with
measures contained in the Crofting Act 2010, they puncture, if not
remove, the threat of speculation in the housing market.

The measures signify a visual disruption of social and spatial boundaries whose ordering used to be defined through the ownership model of
property. A commoning of property rights replaces the practised fixity
and boundedness of that model with fluidity and movement. The measures now exercised by community trusts with respect to property, and
specifically housing, are the outcome of a sometimes lengthy process of
deliberation and negotiation – both within the locale and, varyingly, in
the context of a network of community trusts. The rights to property
with community land ownership are complex and contingent; the
outcome in any particular situation may be uncertain. As I have indicated,
crofters occupy the potentially contradictory subject positions of tenants
with use rights to land under crofting tenure and owners, together with
non-crofters, of the entire estate. Their release of land for housing plots
may reflect the exigencies of very local circumstance. In this counterhegemonic reading of property, rights are constantly in process, their
performance bound up with people's reworking of their individual and
collective subject positions.

Energizing places

I discussed the significance of the generation of renewable energy
through the wind for the sustainability of community-centric economies
in the previous chapter, demonstrating how the right to capture the wind
is tied to the working of property rights. However, in the case of the
North Harris Trust which, after the investment of enormous effort, took
the decision not to pursue the installation of a wind farm at Monan, that

trust turned its attention to the generation of electricity from water. The Assynt Crofters' Trust and the Knoydart Foundation provided precedents. Like the wind, water is a divisible resource whose commodification has the potential to support a community ethic as well as reducing the community's carbon footprint.

Of a number of possible sites identified through a Highlands and Islands Community Energy Company Survey, *The Western Isles Hydropower Feasibility Study* (Faber Maunsell, 2008), the NHT has pursued the site at Bun Abhainn Eadarra. The North Harris Trading Company submitted an application for planning permission for a 150 kW "run of river" hydro-electric scheme for the Eadarra River to the Comhairle in May 2010 and approval was given in August 2010. An invitation to tender was sent out on 1 October 2010, at the same time that this process was initiated by the West Harris Crofting Trust. The two community trusts are in the process of considering offering the work to one company, hoping thereby to reap the benefits of some economy of scale (interview, NHT Energy Officer, 22 March 2011). It is anticipated that the scheme (now expected to generate 100 kW electricity), while not generating as much electricity as the site at Monan could have done, will provide a similar level of income under the terms of the Feed-in Tariff (interview, NHT Energy Officer, 22 March 2011).[15] In addition to the scheme at Bun Abhainn Eadarra, the North Harris Trading Company is investigating the potential for the generation of electricity from micro hydro sites.

The wind is not completely out of the picture. A 10 kW turbine was installed at the Urgha Community Recycling Site in November 2010, supplying electricity to the site and exporting surplus to the grid when conditions allow. Demonstrating the potential that may be released from collaboration between a land owning community trust and a community energy company, the trading wing of Community Energy Scotland (CES-T) has proposed the siting of wind turbines on community owned land. Two 5 kW turbines are planned for Scaladale, to be located behind the Lewis and Harris Youth Club Association Centre, two of a similar size next to the primary school and one for the Croileagan (the kindergarten), both in West Tairbeart. CES-T will own the turbines, receive income from the Feed-in Tariff and sell the electricity at a reduced rate to the consumer. In the case of the Scaladale Centre, the NHT will split income from the rent of the site with the common grazings committee. As the land in West Tairbeart is not part of a common grazings but belongs to the NHT as an "agricultural park", all rental income will accrue to the NHT (NHT Energy Officer, 22 March 2011). As Donnie Mackay, a director, noted in a meeting of the North Harris Trading Company (9 November 2010), it was "a good model" for community land owning trusts: it addressed issues of sustainability and carbon reduction, as well as income generation.

Paralleling the actions of other community trusts, the North Harris Trading Company has additionally taken on the carbon challenge in three other ways. Each project has been designed to decrease the community's carbon footprint. The first involved a partnership between the trust and The Energy Advisory Service and the launch of the Community Carbon Challenge project. The project's objective was to improve the energy efficiency of homes in North Harris, to eliminate fuel poverty and to increase community awareness of energy matters. It offered a free assessment of house insulation needs and then each household was provided with loft and/or wall cavity insulation, without cost, as necessary. By the end of 2010, 101 homes had received insulation through the programme. Home Energy Packs, which include an energy monitor, low-energy light bulbs and PowerDown plugs, were supplied to 300 homes. The NHT contributed to the cost of the project – £10 000 out of a total budget of £126 000. "If you can contribute, [it's] a far healthier outlook than [if you] keep asking for money", i.e. it is "not a begging bowl", said one of the North Harris Trading Company directors (interview, 19 January 2010). If, with all the delays pertaining to the proposal for a wind farm at Monan and its eventual dismissal, the material benefits of community ownership were not very visible to the estate's residents, this project was a visible reminder of what a community trust could do.

A second initiative that increased the local profile of the trust by taking on the carbon challenge concerned the Urgha Community Recycling Site, opened in 2009. Creating a permanent position for an NHT employee, the site acts both as a place to which people bring items for recycling and a site where "re-homing" of items is coordinated. Garden waste is collected and sent to the Comhairle's bio-digester at a site near Stornoway, where it is used to produce electricity and compost; metal goes to the mainland for re-processing. Bicycles are "re-homed" locally or go to a centre in Stornoway for use in bicycle maintenance instruction, televisions find new owners, and, in one instance, slates from a pool table have a new life as a hearth in a North Harris House (*Caraidean*, issue 9, August 2010: 4). The project is popular and considered to be hugely successful. It brings in money by providing a service for the Comhairle (used to pay the salary of the NHT employee), through the sale of electricity from the wind turbine to the grid and, when supplies are available, through the sale of biofuel produced from waste cooking oil obtained from local hotels, the school canteen and "Big D's", the Tairbeart chip shop. The biofuel is used primarily by the NHT's two vehicles.

Third, at Chaolas, in a pilot biomass project carried out by the North Harris Trust in conjunction with parallel schemes with a crofter in Benbecula and with the Laxay Common Grazings Committee in Lewis, willows, together with alder, poplar and sycamore, have been planted to assess their potential as a source of fuel in the particular environment of the Outer Hebrides. At Chaolas, saplings have been planted on two and a half hectares of croft land belonging to three crofters (interview, NHT Energy Officer, 22 March 2011).

The purpose in drawing attention to these initiatives to reduce the carbon footprint is to demonstrate that, once again, a reworking of property – and nature – with community ownership opens up the land to new possibilities. Whether through the capture of wind or water, the sale of electricity to the national grid by a community owned wind farm complicates the process of commodification and supports a community-centred economy. Other measures that are designed to reduce the carbon footprint create a more diverse economy than that based on market relations. The Urgha Community Recycling Site is a case in point. "Waste" is redefined and goods recirculated through alternative and non-market relations (see Gibson-Graham, 2006: 76). Public meetings to discuss the options for renewable energy and other measures to reduce the carbon footprint as well as activities at the Urgha Community Recycling Site create spaces where the meanings of community and a community economy are reworked – where an ethic of collective responsibility is negotiated. Cast as the everyday though they may be, they contribute nonetheless to the strengthening of a counterhegemonic reading of both community and place. They make visible a different working of the land to that defined through private property.

Feeding places

The North Harris Trust's recent initiative in supporting locally grown food through Crofting Connections provides a further example of how the meanings of the land may be opened up with community ownership, albeit as yet in a very preliminary way. With little visible production of food, beyond the ubiquitous sheep,[16] the NHT's enrolment in this venture holds the potential of contributing further to a reduction of the carbon footprint and the building of a biodiverse and sustainable community economy. I trace the possibilities with which the project is linked discursively, recognizing that these links are as yet embryonic, in order to demonstrate its potential significance in reversing a "scopic regime" (Jay, 1994: 589) of neoliberalization of food production.

Crofting Connections is an initiative of the Scottish Crofting Federation and Soil Association Scotland.[17] Launched in September 2009, the three year project focuses on promoting links between school-age children and their crofting heritage. It is underway in 47 schools in the Highlands and Islands, Orkney, Shetland and Argyll. Sir E. Scott School, Tairbeart, is one of the schools involved. The NHT works with the school to further the project. Thus far, the NHT has been instrumental in identifying a site – two unused polytunnels and adjacent land – and establishing a partnership arrangement between the trust, the school and a crofter, John Murdo Morrison, who has agreed in the first instance to a five year period of lease of the land at no cost. The project focuses initially on re-skinning one polytunnel on the site for the use of school children. The NHT Ranger, responsible for the project, anticipates subsequent re-skinning of the second, which, together with adjacent land, will promote broader community involvement. For him, Crofting Connections and further community involvement in food production provides a means through which people can re-attach themselves to the land and, through that connection, engage with "the bigger picture of climate change" (interview, 22 March 2011). Although much of his work has to do with the "wild land" on the estate, connecting with the land "doesn't have to do with getting people to the middle of nowhere", he explained (interview, 22 March 2011).

The project both addresses the issue of the intergenerational sustainability of crofting – of concern given the ageing crofting population and the concomitant threat of loss of agro-ecological skills and knowledge – and encourages children to explore how crofting practices such as collective land management and "high nature value farming" might contribute to the resolution of twenty-first century challenges such as climate change (Rodway, 2009: 16). In one case, in Auchindrain, Argyll, youth in three primary schools are involved in working the land following the runrig system – unused for the previous 173 years – learning in the process about how community was practised at that earlier time (Hamilton, 2010: 11). For Pamela Rodway (2009: 16), Crofting Connections co-ordinator, the project encourages participants to think about "re-localising solutions", a message the Scottish Crofting Federation is keen to promote.

Schools' participation in the project is not dependent on community ownership of the land, nor is the expansion of local food production.

However, if the efforts of the Galson Trust are indicative, community ownership can certainly further local production. Here, the project has been progressed as part of the trust's energy plan by the Powerdown officer, a position funded by Community Energy Scotland. Its successes in promoting local food production include the building of three polytunnels at schools on the estate, the establishment of a number of farmers' markets and the running of a series of "Get Growing Workshops" in conjunction with Lewis and Harris Horticultural Producers (LHHP) (interviews, Directors, Galson Trust, 15 December 2008, 29 March 2011; Maclennan, 2011). While there may be far less agricultural potential for the growth of horticultural production in North Harris than in Galson, there is substantial room for expansion. Support for crofting in general had been relatively neglected in the first ten years of community ownership, and horticulture represents one way of changing this (interview, Chair, NHT, 24 March 2011).

The current interest in horticulture, in North Harris as well as Galson, is tied discursively to broader political agendas. In the Outer Hebrides, the Comhairle and Western Isles Enterprise have supported the work of LHHP, which links its stated aim of increasing local food production to wider objectives of sustainable land use, biodiversity, and recycling.[18] As Ruaraidh Ferguson, LHHP coordinator, has expressed it,

> It's not that long ago when the islands were virtually self sufficient in staple crops, although within a totally different economic context, yet today we are at the end of a very long supply chain, with each family generating thousands of food miles on a weekly basis (Ferguson, 2008: 8).

For Donald Murdie of the Scottish Crofting Federation, it is precisely because "remote places [are] at the end of the longest supply chains [that] ... the greatest opportunities [are found]" (interview, 1 May 2008). "The further you go", he remarked', "the more this [local food production] is a suppressed demand" (interview, 1 May 2008). LHHP's support to growers takes the form of training sessions, the organization of farmers' markets in Stornoway and Tairbeart, and the co-operative purchase in bulk of seeds, other agricultural inputs or machinery.[19]

Through an award from the Scottish Government's Climate Challenge Fund in the autumn of 2009, the LHPP is connected with national and global agendas. The "Get Growing Campaign", as it is called, began in March 2010 and runs for a year, with the option to apply for one further year of funding. Its objective is to expand horticultural production through 200 new growers, thus not only addressing the high demand for locally produced food, but also contributing to a reduction in carbon dioxide emissions, the use of chemicals in agricultural production and the need for excessive packaging.[20] The LHHP has now approached three community trusts – North Harris, West Harris and Galson – as a way of extending its reach.

With its participation in Crofting Connections and its possible future collaboration with the LHHP, the NHT is linked discursively to the movement for food democracy, whose focus on localization offers a space where new political and ecological alternatives to corporate control of the "foodscape" are opened up (Johnston *et al.*, 2009: 514). The broader movement aims to achieve food security and food sovereignty through a focus on citizen rather than corporate control of food production and distribution (Johnston *et al.*, 2009: 515). As is evident in the LHHP's links with other community-led groups (Skye Food Link is one example), with the erstwhile Highlands and Islands Local Food Network,[21] and with the Scottish Crofting Federation's activities *vis à vis* Europe and its membership in La Vía Campesina, the "local" practised by these organizations suggests a reflexive politics where the terms of engagement with the neoliberal model of globalization are reworked. This politics of food localization may be "rooted in place, but [it] simultaneously looks outward to establish solidarity and equality translocally and even transnationally" (Johnston *et al.*, 2009: 527). It contrasts with a politics of "defensive localization" (Hinrichs, 2003: 37) or of "normative localism" (DuPuis and Goodman, 2005: 359), in which "emplacement" is assumed to be "intrinsically more just" (DuPuis and Goodman, 2005: 364) – where a socially homogeneous and coherent local is pitted against the global "other". "A reflexive politics of localism" (DuPuis and Goodman, 2005) demands that the politics of the localization of food be interrogated. Does the process play into the hands of the local elite? Does it provide a means through which the neoliberal project is furthered, i.e. is it easily co-opted by global corporate interests, or does it reflect new forms of governance (DuPuis and Goodman, 2005: 365, 367)? Does the process contribute to an alternative more socially and ecologically just configuration of the global?

In order to develop this argument – and to connect the discussion of the NHT's support of local food production with global discourses of sustainability, biodiversity and social and environmental justice – I turn briefly to consider the significance of the Scottish Crofting Federation's membership in La Vía Campesina, an international movement of small-scale farmers formed in 1993. In October 2009, the Crofting Federation became the 23rd European member (*The Crofter*, 2009: 9). With a membership in 2007 of 148 farm organizations from 69 countries in Europe, Africa, North and South America and Asia (http://viacampesina.org/en/ accessed 13 January 2011), La Vía Campesina has become one of the most prominent and radical social movements supporting an alternative to the neoliberal model of agriculture endorsed by the World Trade Organisation (Desmarais, 2008: 138). In place of this model, La Vía Campesina aims

> to develop solidarity and unity among small farmer organisations in order to promote gender parity and social justice in fair economic relations; the preservation of land, water, seeds and other natural resources; food sovereignty; sustainable agricultural production based on small and medium-sized producers (http://viacampesina.org/en/ accessed 13 January 2011).

At least for the international leadership – the movement is marked by considerable differences among its members – it is the idea of food sovereignty that is central to its vision of a sustainable agriculture and to its response to a globally sanctioned model of agricultural commodification and a "trade-based" model of food security (Wittman, 2009: 813). Food sovereignty, as defined by La Vía Campesina at the World Food Summit in Rome 1996, includes the right of peoples and countries "to define their agricultural and food policy", to organize "food production and consumption [to meet] the needs of local communities, giving priority to production for local consumption", and to ensure that "[l]andless people, peasants and small farmers [have secure] access to land, water and seed as well as productive resources and adequate public services" (http://viacampesina.org/en/ accessed 13 January 2011).

As evident in the ritual exchange of seeds in their own meetings (Desmarais, 2008: 141) as well as in their participation in various meetings of the Convention on Biological Diversity and elsewhere, seed sovereignty is integral to La Vía Campesina's conceptualization of food sovereignty (Wittman, 2009: 817). The organization protests both the concentration of seed production in the hands of a few multinational corporations and the genetic modification of seeds, particularly the "terminator" or "suicide" seeds, which are coded to prevent their reproduction and thus farmers' ability to save seed from one season to the next (Wittman, 2009: 817). "The dissemination and preservation of locally adapted seeds", writes Hannah Wittman (2009: 818), "even in incipient and localized ways, thus could begin to disrupt the trend towards specialization and commoditization within the dominant model of agricultural production". They are taken by La Vía Campesina to "ground" debate, a material and metaphoric reminder of the integral connection between "peasant" farmers and the land (Wittman, 2009: 818).

In light of this, the growing visibility (literally and figuratively) of indigenous species and older varieties of contemporary crops on farmers' fields in the Outer Hebrides and the LHHP's connections with the Orkney Agronomy Institute (*LHHP Newsletter*, November 2005: 3) open up interesting questions. One points towards the possibility that seeds may provide a means through which the local may be reconstructed in these islands – a local that is more genetically diverse and agro-ecologically complicated than is the present norm. Bere (*Hordeum vulgare*), closely related to present varieties of barley, is central to the Agronomy Institute's research agenda and to the disruption of this norm (*LHHP Newsletter*, November 2005: 3). Cultivated most widely in Orkney, but with a presence in the Uists, bere flour has been used for hundreds of years for making bannock (a round, flat, unleavened bread baked on a griddle), bread and biscuits. In Orkney as well as elsewhere in Scotland, bere has also long been used for malt (for whisky) and beer. Valhalla Brewery in Unst, Shetland, continues this tradition, as do homebrewers in Orkney (Martin and Chang, 2007: 29). The Agronomy Institute is now exploring the possibility of extending the present geographical range of bere and of expanding the number of products for niche-marketing, recognizing the uniqueness of bere and its strong local identification in the Highlands and Islands (Martin and Chang, 2007: 29).[22,23]

Maria Scholten's research into the "small oat" (*Avena strigosa*), or *Corc beag* as it is known in Gaelic, integrated into the new National Crofting Course at Sgoil Lionacleit, Benbecula's high school, provides a further example of how seeds are re-attaching people to place (interview, 10 June 2010; Scholten *et al.*, 2010). Supported by the Scottish Agricultural College in 2009 and 2010, Scholten's research alerts students in a practical way to the critical role that crofters play in genetic diversity, crop breeding and *in situ* conservation on the machair, producing seeds that are closely adapted to particular ecological niches. Corc beag, an ancient and distinct species of oat, a landrace, is now rare.[24] Previously widespread in Europe and elsewhere in the UK, its main stronghold is now in the Uists and Benbecula, where it is grown on about 400 hectares (Scholten *et al.*, 2008: 5). It is also grown in Shetland – the Shetland oat or "ate" – and on Tiree – the Tiree oat (http://species. bsbi.org.uk/html/avena_strigosa.html accessed 18 November 2009). Its more recent decline – from about 1400 ha in the 1980s to about 400 ha in the early twenty-first century – particularly on Lewis and Harris and, to a lesser degree on North Uist, is due to a declining and ageing crofter population, decreasing resilience in seed production owing to the smaller numbers of seed producers, the replacement of cattle by sheep and the growing use of silage (Scholten *et al.*, 2008: 3, 5). Both the decline in seed production and the increased use of silage are linked to the threat of crop damage by greylag geese, a protected species, whose numbers have doubled over the past decade (Scholten *et al.*, 2008: 3, 5). As Scholten (2008: 18) has shown, the small oat used to be multiply cropped with bere and rye, a combination that provided a measure of insurance against crop failure in a difficult environment. Corc beag survived in the islands, she writes (Scholten, 2008: 18), as public plant breeding no longer focused on "marginal environments" and thus crofters had to rely on their own time-honoured skills of seed selection. "There is a rich source of local knowledge, insights and experience to tap into", she continues, "and a tradition of resilience and independence" (Scholten, 2008: 18).[25]

The seeds' significance, in terms of the argument I am making here, is that they provide a further means through which people re-create place. Collectively, through the production of seeds that provide a common good, crofters bind themselves to a place that is more securely genetically diverse. Discursively, this place is connected with other places in Scotland where seeds are regrown by crofters – *in situ* conservation – and to the site near Edinburgh Airport where, close to the Scottish Agricultural Science Agency, Scotland's collection of crop varieties is held. This collection includes landraces such as bere which have been brought here under the Scottish Landrace Protection Scheme established in 2006 – *ex situ* conservation (Barbor-Might, 2008). George Campbell, vegetables trials' manager, interviewed by Dick Barbor-Might (2008), referred to the "'patriotic' value of landraces", by which he meant "the intense pride crofters have in locality – the island landscapes in which the crops grow, the people there and the colours, textures and shapes of the vegetables and cereals". Kevin O'Donnell, head of rural scientific services, remarked that, "Scotland's cultural history is not just castles and works of art. There is also our biological heritage in the form of these ancient crop varieties" (cited by Barbor-Might, 2008).

The localization of food production and support for plant genetic diversity does not depend on community ownership of the land. However, community endorsement of local food production, as is the case in Galson and is incipient in North Harris, has a significant role to play in the politics through which such food and seeds are secured. To refer back to questions raised concerning food democracy, there is, I would argue, less likelihood of the politics of food production being co-opted by particular interest groups where the initiative is led by a democratically elected community trust on land that is collectively owned than where this is not the case. In other words, such an initiative is more likely to produce a questioning of the process of neoliberalization of agricultural production.

Narrating places

As a final instance of the North Harris Trust's reworking the land in the collective interest, I focus on the trust's support of archaeological investigation and rehabilitating the network of paths. Both further the production of a counter-visuality of place to that of a land empty of people. Both sets of activities are about what Edward Said (1994: 226) calls a "culture of resistance" – "reclaim[ing], renam[ing], and reinhabit[ing] the land". They produce a land which has been peopled through the millennia, troubling an imagined boundary between the social and the natural.

To turn first to the archaeological record, compelling evidence of the historical depth of the bond between people and the land/nature comes from three studies initiated by the trust and its trading wing, the North Harris Trading Company, not long after the land was brought into community ownership. One, commissioned by the NHT, related to the proposed planting of native woodland in Gleann Langadal (Chapter 3) (McHardy, 2006b). A second, commissioned by the North Harris Trading Company and carried out by archaeologist Carol Knott in 2003, was associated with the North Harris Water Quality Programme. The third, undertaken by Knott and Duncan MacPherson, on behalf of the Harris Archaeology Group, took place in 2006 and was occasioned by the planning requirements for the proposed wind farm at Monan (Chapter 4). It is Knott's study of 2003, conducted in seven sites in North Harris, that provides the most detailed evidence of the lengthy period of human occupation of North Harris, and for that reason it is the focus of attention here (Knott, 2003). All three studies relied on a walk-over survey and the analysis of archival sources.

Knott's 2003 study is important. It identifies "extensive and well-preserved" remnants of Scots pine and birch woodland (stumps, birch bark and pine cones) dating back to the Holocene at Aird a'Mhulaidh (Knott, 2003: 14).[26] Using as evidence place names, archaeological investigation and oral history, the study documents early human occupation at Huisinis and Maaruig. The name "Huisinis", itself, is derived from Old Norse, meaning "House on the Headland" (Knott, 2003: 3). And the name *Cill Choinnich*, denoting a site in the Huisinis area, refers to the cell of an early Celtic saint, St. Kenneth (Knott, 2003: 4). At Maaruig, a polished stone axe-head unearthed during construction activities in 1966 is dated to the Neolithic (Knott, 2003: 7). At Maaruig also, one of the burial grounds is "tentatively" linked to a "possible" church dedicated to another Celtic saint, St Rufus or Maelrhubha (Knott, 2003: 6). Other sites provide evidence of more recent occupation – of pre-Clearance settlements and associated feannagan – a land that was closely peopled and deeply worked.

While pre-nineteenth-century archaeological evidence in North Harris does exist outwith the study area – the Both a'Chlair Bhig beehive shielings on the banks of the Abhainn a'Chlair Bhig not far from Loch Resort in the far interior of the estate providing one example (see Burgess, 2008: 86) – Knott suggests that their relative paucity reflects the need for further research rather than the absence of such material. Aware of this situation, the NHT has actively encouraged further archaeological investigation. John Hunter, professor, together with students from Birmingham University and local volunteers, has carried out a walkover survey of North Harris, the findings relayed to the Western Isles archaeologist for inclusion in mappings of the island's archaeology. It is to this initiative that credit is given to making the estate's archaeology much more visible (NHT, 2007: 22). In the conduct of the research, significant emphasis was placed on involving local people, assisted in this case by an active local archaeological group, Linn gu Linn (From Age to Age) (see Colls et al., 2010). Through this group, local ownership of the North Harris Estate's and, more broadly, the island's archaeological record was ensured. Outreach activities at local schools furthered the process, students participating in classroom activities as well as on-site visits. Interpretive leaflets to be produced to accompany the footpaths that I discuss shortly will circulate this knowledge even more widely.

The NHT has furthered the visibility of the archaeological record and local embeddedness of the knowledge so produced through support for the Whaling Station Action Group, formed, under a different name, over 20 years ago, with a remit to progress the conservation and development of the whaling station at Bun Abhainn Eadarra.[27] The group comprises interested individuals and representation from Harris Development Limited. Since 2003, the initiative has been linked to the ambitions of the NHT, which now owns the site and has identified the whaling station as a priority in its Business Plan 2007–2018 (CIB Services, 2008a: 15). The project has substantial local support on account of its historical significance – culturally, socially, and economically. Named a Scheduled Ancient Monument in 1992, the whaling station is referred to as

Photograph 5.2   The Whaling Station at Bun Abhainn Eadarra.
Source: The North Harris Trust. Used by kind permission.

"the finest and best preserved example of a shore-based whaling station in the UK" by the consultants commissioned by the NHT to produce a conservation management plan (Headland Archaeology and John Renfrew Architects, 2007: section 2.1.1). On account of its links to Norway and the global paucity of remains of such stations, the consultants consider it of world-wide importance (Headland Archaeology and John Renfrew Architects, 2007: section 4.3.2). "[I]t offer[s] a unique example of both international and local involvement in the whaling industry in the Western Isles during the first half of the 20th century", they write (Headland Archaeology and John Renfrew Architects, 2007:

section 4.3.2). The remains provide evidence of the three phases of the whaling station's history and comprise an extensive complex of buildings, among which the intact tall brick chimney is visually dominant in a site that, as a whole, has "a strong visual presence and is highly atmospheric" (Headland Archaeology and John Renfrew Architects, 2007: section 4.3.10) (Photograph 5.2).

The whaling station was established by two brothers, Carl and Peter Harloffsen, from Tonsberg, Norway, through the transferral of the Harpunen Company from Iceland in 1904 (Headland Archaeology and John Renfrew Architects, 2007: section 3.6.1). Operations began in that year and lasted until 1928 with an interval during World War I. In 1922, following the decline of the Norwegian company, the Harris Whaling and Fishing Company, a venture of the Lever Brothers, took over operations. The co-founder of this company was Lord Leverhulme, who had bought Harris and Lewis in 1918 (Hutchinson, 2003). This company tried to expand the business in a number of ways, including experimenting with the export of whale meat sausages to Africa, an initiative that ended in failure (Headland Archaeology and John Renfrew Architects, 2007: sections 3.6.3, 3.6.4). Operations ceased three years after Leverhulme's death in 1925 as a result of a decline in whale stocks (Headland Archaeology and John Renfrew Architects, 2007: section 3.6.5). They were rekindled briefly at the beginning of the 1950s by Captain Jesperson, a Norwegian, ceasing in 1953 due to their unprofitability (Headland Archaeology and John Renfrew Architects, 2007: section 3.6.7). At their height, the industry employed 158 women and men, that number dropping to 50 in the 1950s, a figure that excludes the Norwegian crew operating the catcher, as the boat was called (Headland Archaeology and John Renfrew Architects, 2007: section 3.6.6).

With community ownership of the land, archaeological investigation, as practice, has become a means through which people remake place. The maps, photographs and action on the ground make visible a compli-cated land, where people's stories have been interwoven with the land through time: from the Mesolithic hunters of 8000–4000 BC in the immediate post-glacial period, the Neolithic farmers of 4000–1800 BC and the bronze age metal workers of 1800–800 BC to the Iron Age (800 BC–800 AD, during which time [600–800 AD] Celtic Christianity was introduced in the islands), the arrival of the Norse in the early ninth century and their influence through what has been called The Viking Age to 1266,[28] the medieval period, the beginnings of crofting, the Clearances of the nineteenth century and the present (Burgess, 2008). This is a rich and complex history. It provides evidence of a land/nature worked for millennia by people who were part and parcel of contestations of power from further afield (see Armit, 1996). This mapping of the land counters the production of North Harris as a place divorced from the cross-currents of history – as suggested by those whose wilding of the land is based on class interest or on the protestations of less social conservationists. It provides one of the means whereby the community owned estate creates a different visuality, where the interweaving of the social and the natural are evident.

The restoration of a 35 mile network of footpaths and the production of interpretive material – signage, leaflets, a website, some still at the planning stages, which guide the participant through the diverse socio-natural landscape – further this counter-visuality.[29] The paths are a reminder of the historical depth of the peopling of the island. Some connect crofting townships; others were built for the purpose of stalking at a time when that "sport" provided the *raison d'être* of the estate. Others still were coffin routes, necessitated by the restriction of burial grounds to the deeper soil of the machair in the west. One of the more dramatic paths connects Urgha to Reinigeadal. Its well-engineered zigzags over the *Scriob* provided the route between the two townships, used twice a week for the delivery of the post. It and other paths are highlighted on the map printed on one side of the high-quality, publicly available (free) leaflet, *Urras Ceann a Tuath na Hearadh. The North Harris Trust*. In addition to the map, the leaflet provides the visitor with information about the trust – "who we are", "what we do" – as well as photographs and brief texts about sites of interest.[30] The visitor is reminded, for instance, that it was not until 1990 that a road to Reinigeadal was completed. Prior to this time, people relied on the footpath – or boat.

The path to Reinigeadal from Urgha has been the subject of substantial remedial work – clearing culverts, building water bars and cross drains, and constructing wooden bridges. Work has also been carried out on the Bogha Glas to Gleann Langadal path and part of the path from Loch Chliostair to Gleann Ulladal.[31] Some of the labour has been carried out by experienced contractors, some by volunteers. The NHT has organized volunteer days for the purpose of both carrying out necessary maintenance work and providing a visible and direct means through which the community could be engaged in trust activities (Calum MacKay, Chair, NHT, cited in *Stornoway Gazette*, 16 February 2006: 19). Each year, the John Muir Trust organizes work parties, which come to the island for a week at a time[32] (Photograph 5.3). The Harris Stalking Club has also provided volunteer labour. And in the case of the rebuilding of the bridge at Scourst in Gleann Mhiabhag in 2005, the NHT drew

Photograph 5.3   Path rehabilitation by John Muir Trust volunteers.
Source: The North Harris Trust. Used by kind permission.

on the labour and expertise of the Royal Engineers with the City of
Edinburgh Universities Officers Training Corps. In order to increase local
capacity, the NHT has promoted the training of local people in footpath
construction and maintenance. In 2005, for example, the National Trust
for Scotland's Footpath Training Team worked alongside trainees for two
weeks on the path stretching west from Bogha Glas (*Stornoway Gazette*,
15 September 2005: 14). The team's expertise lies in construction using
time-honoured methods which minimize environmental impact. The exercise contributed to the trainees' Scottish Vocational Qualification in
Environmental Conservation. More recently, the Land Manager has
looked into the possibility of the delivery of certificated training courses
in footpath construction and management by Lews Castle College or, in
the case of shorter courses, by Western Isles Rural Opportunities (Land
Manager's Report, 25 January 2010).

Maintenance of the paths is not simply about enhancing a visitor's
enjoyment of the landscape – important though tourism is for the island's

economy[33] – or providing jobs for local people. Paralleling the archaeologists' exposure of the land as worked through the millennia, maintaining the paths is part and parcel of the production of a counterhegemonic landscape. The paths' re-inscription on the land recalls stories through which the land has been produced through the past hundreds of years. Some paths have been created and used for the pursuit of stalking; most bring to mind the quotidian practices of people as they work the land and reforge the bonds of community. Once again, they suggest a troubling of an imagined boundary between the social and the natural.

Guided walks by the North Harris ranger, an employee of the NHT, carry a similar message. Created in 2009 with financial assistance from SNH and the Esmee Fairburn Foundation, the North Harris Ranger Service is responsible for working with the public – local people and visitors – with a view to "increasing awareness, understanding, care and responsible use of the natural and cultural heritage of North Harris" and furthering "the sustainable management and use of the outdoors" (http://www.north-harris.org/what-we-do/land-management/education-and-interpretation). In the guided walks, often using the footpaths as part of the route, the focus may be the machair, a stretch of rocky shore or an eagle high in the hills. However, walkers are constantly reminded by the ranger of the inter-working of the social and the natural and of what it means to be part of a land owning community trust. The walks are advertised in a pamphlet containing information about walks sponsored also by Harris Development Limited and by the Galson Estate Trust, its images evocative of a "wild" land – a golden eagle, a corncrake, deer, an orchid, snow-capped hills, stretches of white shell-sand beach – but bound to past and present human action through a trigonometrical point at the summit of a hill or the now unused feannagan of a crofting township.[34] The walks and the photographs are, in the NHT ranger's words, a reminder that "I'm not just a visitor in this landscape, but I'm part of it" (interview, 22 March 2011).

The broadcast of a live climb of a new route on Sron Ulladail, North Harris, the UK's longest overhanging cliff, by Dave MacLeod and Tim Emmett, rated among the world's best climbers, emphasizes the point. The event, filmed on 28 August 2010 by Triple Echo Productions for the BBC, was the focus of a five hour programme that included footage elsewhere on North Harris, including details about the community trust. Of Richard Else, producer, the NHT Land Manager said that he "loves mountains but wants communities in them" (interview, 20 February 2010).

A Mountain Festival, held in September 2011, built on publicity generated by the Great Climb. Events highlighted the excellence of hillwalking in North Harris. There may be no Munros – so-called after Hugh Munro who, in 1891, published his first tables of all mountains in Scotland 3000 feet or more in height – but An Cliseam (799 m) and other hills "make up for their modest height by the grandeur of their setting, their magnificent views, and their abundant wildlife", remarks the NHT's Land Management Plan (2007: 20). Festival events included an address, classes and photography of the North Harris Estate by Laurie Campbell, Scotland's leading nature photographer. The events were linked to the building of a golden eagle viewpoint in Gleann Miabhag, a nature trail at Huisinis together with the production of an interpretive leaflet and, on 1 October, a community walk from Miabhag to Bogha Ghlas with mini buses at the far end to bring people back to their cars (interview, NHT Land Manager, 28 March 2011).

What I am arguing here is that, through activities such as archaeological investigation, the rehabilitation of footpaths, the ranger's guided walks and events such as the live climb on Sron Ulladail and a Mountain Festival, a counternarrative or counter-visuality of the estate is created. Paths that lead through currently uninhabited parts of the estate, climbing and hillwalking may evoke images of "wildness" and "remoteness" (NHT, 2007: 20) – the subject of the tourists' gaze – but they also make visible community interest. This is no longer an estate where nature as commodity is the exclusive purview of the wealthy or where it is the subject of external conservationist interests. Through local people's participation in the activities I have detailed and the un-earthing (literally) of evidence of a land of which humans have for millennia been part, a norming of nature which had previously provided the scaffolding for both class and conservationist politics is "undone". "Nature" and the "wild" are opened to different meanings, their ordering as historically inert ontological categories suspect.

Placing community

In this chapter, I have explored how a-commoning of property rights through community ownership has opened the land to new imaginings. These I have traced through specific projects. In and of themselves they may appear small – the stuff of the everyday. However, together they are beginning to demonstrate that there is another way of seeing the estate. The building of affordable housing starts to tackle the problems that have emerged through the marketization of crofts. Initiatives to derive energy from renewable sources and reduce the carbon footprint through a scheme to provide home insulation, the highly popular recycling site, the beginnings of support for local food production and efforts to make visible the community's past through archaeology and the restoration of the network of footpaths contribute to a reversal of the ways in which this place was "*given* to be seen" (Rajchman, 1991: 69, emphasis his).

I could have cited others. There is a Community Development Fund to which groups can apply for particular projects. Support for Gaelic, for instance through the annual Gaelic festival, Feis Eilean Na Hearadh, is one example, giving visibility to an area that the NHT has relatively neglected thus far but that it has identified as a priority in its objectives. Improving the jetty at Amhuinn Suidhe, which has benefitted particularly local fishers, is another. There are plans to build zero carbon business units to attract small businesses to the island and thereby increase employment opportunities[35] as well as a long-standing ambition to build a Harris Tweed centre to showcase that icon of an island's heritage. The NHT is now itself a major employer. The seven employees comprise an administrator and assistant administrator, a development manager, a land manager, an energy officer, a ranger and a recycling assistant.

Each initiative provides the political space for people to rework their individual subject positions *vis à vis* the collectivity. The projects may be initiated by the NHT leadership, but, in their execution, the NHT actively engages with its members on the grounds of both principle and practicality. I am not trying to suggest that there is never controversy, but there is repeated evidence of a politics of inclusiveness. In some cases the reconfiguration of the community has been based on extensive consultation, as was the case of the proposed wind farm at Monan. In 2005, a detailed questionnaire, "Your Opinions Count", was distributed to members, its objective being to canvas people's ideas and priorities for trust action. In the case of footpaths, residents on the estate provide volunteer labour, as do the annual John Muir Trust work parties. As Calum MacKay, Chair, NHT, commented with reference to a specific volunteer day, 4 February 2006, called to repair the walkway in Gleann Miabhaig, such events are well supported. On this occasion, 30 people turned out (cited in *Dè Tha Dol?*, 2006: 9). Members of the committee that adjudicates applications for the Community Development Fund are drawn from the trust's membership. Members of the community with particular skills are invited to sit on the board of the North Harris Trading Company. And it is members who organize the annual ceilidh and Tiorgha Mhòr, the hill race. Youth are now more directly involved than they were in the past through Crofting

Connections and through tree planting on land adjacent to Sir E. Scott School. Members are kept informed about trust activities on an ongoing basis through *Dè Tha Dol?*, the bi-weekly newsletter produced by Harris Voluntary Service, circulated throughout the island and, on occasion, by press releases and mail.

What I am trying to convey here is that, through bringing land into collective ownership, a whole new set of actions has been generated. The land and the people on it are, literally and figuratively, a-working. The actions provide the means through which individuals reposition themselves with respect to the community and, at the same time, produce the beginnings of a new community economy. While "the negotiation of interdependence" that, Gibson-Graham and Roelvink (2010: 330) suggest, lies at the "core" of a community economy has long been practised by the crofting community – through such collective practices as the gathering, shearing and dipping of sheep, together with the management of the common grazings and the allocation of souming and, since 1991, the planting of trees – this new "econo-sociality" (Gibson-Graham and Roelvink, 2010: 330) is now taken forward by the broader, place-based community.

In turn, community and a community economy "become" through the interconnections among land owning community trusts. Examples that I have cited here and in Chapter 4 include workshops on energy, housing and land use funded by Highlands and Islands Enterprise's Community Land Unit and HICEC. They also include such initiatives as that of the NHT's invitation to other community land owning trusts in the Outer Hebrides to meet in February 2011 to discuss future collaboration on a range of economic and social matters.[36] These fora provide the political spaces where knowledge is shared and the threads of social and economic interdependence strengthened. And it is this direction, I would suggest, that gives credence to the argument that, while "community" may well be integral to a governmental agenda that seeks to divest itself of responsibility for addressing the negative effects of neoliberal policy, community trust members "can and do" create the political space for a counternarrative. They do this through a discourse of community in ways that parallel the production of "new political subjectivities" through discourses of "multicultural-ism" and "stakeholder participation" of which Katz (2005: 630) writes with reference to the Andes and Aotearoa New Zealand. What I would suggest is incontrovertible evidence for this line of reasoning comes from the formation of Community Land Scotland in 2010, whose significance I now turn to consider.

Notes

1 In 2005, tenants of council houses voted in favour of transferring ownership from Comhairle nan Eilean Siar to Hebridean Housing Partnership, a
   charitable housing association. This move was part of a national initiative
   to pass ownership and management of rental housing to "not-for-profit
   social landlords", in return for which the Comhairle's housing debt of £38
   million was erased and £12 million was provided for new housing (Editorial,
   WHFP, 4 November 2005: 7). The aim of the move was to allow the housing organization to access funding from the private sector, thereby enabling
   it to build more affordable housing than had previously been possible and
   to carry out more extensive improvements of the existing housing stock
   (Editorial, *WHFP*, 4 November 2005: 7). Tighean Innse Galle is Hebridean
   Housing Partnership's development agent.
2 For information on the NHT's position with respect to housing, I am
   indebted to the NHT's Development Manager.
3 Scotland-wide, the number of second and holiday homes increased from
   19 756 to 29 299 between 1981 and 2001 (Mackinnon, 2005: 12).There are
   no available statistics on the number of holiday homes in North Harris. The
   West Harris Crofting Trust records that 41 per cent of the houses on their
   land are holiday homes (CIB Services, 2008b: 2), a figure that is likely to be
   higher than the percentage for North Harris. The West Harris Crofting
   Trust's figure is matched by the Inner Hebridean islands of Colonsay and
   Iona, both of which have 40 per cent of their housing stock classified in this
   way (Rural Housing Service, 2002: 7). On Gigha, in contrast, where 68 per
   cent of all housing is owned by the Gigha Heritage Trust, only 10 per cent of
   the housing is holiday accommodation (Rural Housing Service, 2002: 7).
   Properties owned by the Gigha Heritage Trust are rented out on a variety of
   different terms (Rural Housing Service, 2002: 8). In an effort to curtail the
   trend towards holiday homes, in 2008, Comhairle nan Eilean Siar decreased
   from 50 to 10 per cent the discount on council tax for second homes or
   houses that were empty for much of the year (MacInnes, 2008c: 11). Income
   raised from this source (estimated at about £300 000 per annum) is re-
   invested in the provision of affordable housing under the Additional Council
   Tax Income from Second Homes programme (MacInnes, 2008c: 11; *WHFP*,
   18 April 2008: 9).
4 In a study of rural gentrification in five sites in Scotland, Stockdale (2010)
   cautions against viewing all in-migrants in undifferentiated terms. Her
   argument, with respect to housing, is that the greatest likelihood for the
   displacement of local people is found where the greatest income differential
   exists between the income levels of the local population and those of the
   in-migrants (Stockdale, 2010: 35).

5   Ending the Right To Buy on "new-build" affordable housing is viewed as
    one of the significant measures introduced by the Housing (Scotland) Act
    1991 to ensure the long-term availability of such housing stock
    (Campbell, 2009: 15). Scotland-wide, the stock of social housing
    decreased from 691 524 homes in 2001 to 598 922 in 2008 (Campbell,
    2009: 15), a decline attributed in large part to the exercise of the Right
    To Buy introduced under Margaret Thatcher's government in the1980s
    (Campbell, 2009: 15).
6   One frequently cited reason given for the lack of house construction
    outwith Stornoway is the extra cost. In the Outer Hebrides, it is estimated
    that it costs 10 per cent more to build a house in rural areas compared with
    Stornoway (WHFP, 8 May 2009: 1).
7   Angus MacLeod also pointed out that they were in the process of putting
    in place measures that would ensure, in the event of a subsequent sale, that
    there was some repayment of the money the purchasers saved on buying
    the land at below the market price (WHFP, 21 October 2005: 3).
8   As an example, decrofted house sites were reportedly being sold for £25 000
    in parts of Lewis in 2004 (WHFP, 3 September 2004: 3). In some cases, a
    crofter would apply for planning permission for multiple sites on one croft.
    One prominent case occurred in Aignish, Lewis, where an absentee crofter
    applied for planning permission for four houses on his croft. This action led
    the Aignish Common Grazings Clerk, Donnie MacDonald, to remark
    that, until recently, Aignish consisted of 32 houses on 32 crofts. Now the
    number was close to 80 houses. "Despite the township's objections over the
    years", he commented, "the [Crofters] Commission has consistently – and, it
    could be argued, rather arrogantly – ignored the wishes of the community"
    (cited in WHFP, 22 September 2006: 3).
9   Logie (2007: 5) gives a further example from Kilvaxter, North Skye, where
    an individual who had benefited from a grant from the Croft Entrant
    Scheme in 2004 was, in 2007, offering the tenancy of his croft, excluding
    the house, for £22 000, and putting up for sale an additional three house
    sites (occupying 1.2 acres) with planning permission for a combined figure
    of £149 000, together totalling £171 000. (Also, see K. Mackenzie, 2007.)
    In the crofting township of Horgobost, South Harris, a house was placed
    on the market in 2006 for the "astonishing price of offers over £350 000"
    (WHFP, 10 February 2006: 3). Local resentment ran high as the house had
    been sold to the present owners two years previously by a local crofter for
    a "modest price to allow a young couple to set up home in the area" (WHFP,
    10 February 2006: 3). The sellers claimed justification for their asking price
    on the grounds of improvements to the property.
10 Surprisingly, given the committee's strong support for the devolution of
    power to communities (Committee of Inquiry, 2008: 57), the committee
    neglected to consider the role that community land ownership might

play in the process of resolving the issue of housing in areas under crofting tenure.

11   The Scottish Government's (2008c: 15) rationale was as follows:

> The Government believes that it is only necessary to consider an occupancy condition where land has been decrofted (or resumed by the landlord). Such an occupancy condition would require any house on land that has been decrofted, or that is subsequently built on land that is decrofted, to be used as a permanent place of residence. The enforcement of this condition would not be a matter for the crofting regulator, but for the relevant local authority.

12   The process of becoming a Rural Housing Body involves an application to the Scottish Government. Their decision is based on an examination of a trust's constitution (Young, 2004: 7). The Stornoway Trust's application for this status was refused on the grounds that providing housing was not identified in its constitution (Logie, 2007: 35).

13   An alternative, adopted by the Highlands Small Communities Housing Trust, which has made substantial use of the Rural Housing Burden, is to calculate the buy-back price as a form of shared equity (Logie, 2007: 35). By this means, the housing trust seeks to balance the maintenance of affordable housing – a community interest – with an individual owner's interest in having a fair return on their investment (Logie, 2007: 35).

14   Recognized as a "Local Letting" Initiative in the Housing (Scotland) Act 2001, the allocations policy allows a community (or housing association or Registered Social Landlord) to prioritize its own needs (D. Alexander, Highland Small Communities Housing Trust, speaking at the Community Housing Conference, Gigha, 2 November 2006). For example, if there is a need for young people in order to keep a school open, this can be reflected in the points system (D. McDonald, speaking at the Community Housing Conference, Gigha, 2 November 2006).

15   The Energy officer explained to me that, although planning permission has been granted for a 150 kW installation in North Harris, there is an incentive to downsize to 100 kW as the Scottish Government's Feed-in Tariff is almost double for smaller turbines (interview, 8 November 2010). A grid connection has been secured for January 2012.

16   Changes in the support mechanism for livestock from one based on headage to one based on area under reforms to the Common Agricultural Policy have led to an estimated drop in the number of sheep by 60 per cent between 1999 and 2006 on the North Harris Estate (NHT, 2007: 19). The NHT's *Land Management Plan* (2007) considers that the decline will continue after 2006, when the introduction of the Less Favoured Area Support Scheme "removes the need for a minimum stocking density to claim full payment" (NHT, 2007: 19). The plan comments: "The situation may offer

some short term respite for overgrazed grassland and moorland, but the loss of hefted flocks from the Harris Mountains will be a lost management resource in the long term" (NHT, 2007: 19).

17 As an organization that defines itself as "the only member-led organization dedicated to the promotion of crofting and the largest association of small-scale food producers in the UK" – the caption that appears at the top of each edition of *The Crofter* – the Scottish Crofting Federation's support of crofter food production has grown substantially over the past decade. Promotion of a crofting brand has helped, as has the push towards croft diversification, including into horticulture, evident in the focus of the annual gathering and in the numerous training sessions organized for crofters by Donald Murdie, project co-ordinator for the Federation's Crofting Resources Programme. One example of an annual gathering that focused on food production was that which took place on Barra in 2008. Entitled *Crofting. Delivering the Goods*, it included sessions on "Selling high value nature", "Marketing croft produce", "Highland stock" and "Ethical sourcing". At this meeting, Environment Minister Michael Russell, MSP, launched the Scottish Crofting Produce Mark. A grant received from the Scottish Government's Food Processing, Marketing and Co-operation Grants Scheme in 2009 – to promote new producer groups and collaborative marketing – strengthens further the Scottish Crofting Federation's commitment to an expansion of crofter food production (*WHFP*, 6 March 2009: 15). For a discussion of the beginnings of what was first known as the Scottish Crofters Union, see Hunter (1991).

   Soil Association Scotland is a charitable, membership-based organization that promotes organic farming and is the main body responsible for organic certification in the country.

18 Interest in local food production stemmed from a study carried out in 1996 by West Isles Enterprise, the Comhairle and LEADER II, which focused on the local cultivation of fruit and vegetables and demand for such produce (*Stornoway Gazette*, 14 June 2001: 3; CnES, 2008b). This study was followed by crop trials in 1998 and a subsequent study supported by Western Isles Enterprise and the Scottish Agricultural College into how to market the produce (*Stornoway Gazette*, 14 June 2001: 3). Local produce markets emerged as the preferred choice.

19 Training sessions have included such diverse topics as the basics of fruit and vegetable production in the islands, organic farming, composting and shelter belts (promoting the growing of willows as hedges). They have been delivered by, among others, experts from the Scottish Agricultural College and the Scottish Crofting Federation. The market in Stornoway, started in 2000, now runs throughout the year on Saturdays and in the summer months also on Fridays. Reflecting the smaller population base of Harris, the Tairbeart market, which began in 2006, is held on Saturdays in the summer (*LHHP Newsletter*, January 2007: 7).

20   As before, training is a key part of the campaign; six sessions were held on Harris in 2010, an indicator of the emphasis placed on this aspect of the programme.

21   The Highlands and Islands Local Food Network (HILFN) was created in 2004 with support from Highlands and Islands Enterprise. Its objective was to link producers with consumers in a variety of ways in order to promote local food production; it also aimed to build a deeper knowledge and skills base in this geographical area. The network also aimed to promote collective trading with respect to supplying schools and hospitals. In May 2007, the network put in a tender to supply one-third of the Highland Council's food supplies (which included 168 school canteens, 40 care homes and 11 offices) with food of local provenance (from 18 larger farms and 96 crofts and smaller producers) (*WHFP*, 18 May 2007: 19). In support of its bid, the network claimed a number of benefits of "buying local". These included an increase in the number of children eating school meals, a reduction in food miles and in carbon emissions, the retention of about five jobs and the creation of four more, the support of the principles of Fairtrade for local farmers and crofters, and contributing an additional £500000 of economic activity to the local economy (*WHFP*, 18 May 2007: 19). In the event, the Highland Council's decision to award contracts for the procurement of food to large-scale contractors outwith the area rather than to the Highlands and Islands Local Food Network, ostensibly on economic grounds, proved highly controversial (*WHFP*, 10 August 2009: 1, 3). Local procurement would have added 7p per day per school meal (D. Murdie, interview, 1 May 2008). According to D. Murdie (interview, 1 May 2008), although this attempt by the HILFN failed, the Highland Council will now take food quality and provenance into account in its decisions regarding school meals. The HILFN was suspended in March 2009 on account of Highlands and Islands Enterprise's withdrawal of funding.

22   On the basis of archaeological evidence, bere has been dated to the Early Iron Age (Scholten *et al.*, 2010: 4–5). Oats date from the Late Iron Age and rye from its introduction by the Vikings (Scholten *et al.*, 2010: 4–5). The importance of bere may be assessed from historical evidence that indicates that, under the runrig system in Orkney, bere was grown on the most fertile inbye land, whereas oats were grown on more marginal land (http://www.agronomy.uhi.ac.uk/html/Bere_research.htm accessed 16 June 2009).

23   Peter Martin and Xianmin Chang (2007: 29) of the Orkney Agronomy Institute indicate that bere declined with the introduction of higher yielding, short-strawed varieties of barley. For both brewers and distillers, the lower grain nitrogen levels in the new varieties meant a higher alcohol yield (Martin and Chang, 2007: 29). Despite bere's lower alcohol yield, Island Bere, introduced in Orkney in 2006, using bere, has proved to be highly popular (Martin and Chang, 2007: 29). The institute has focused on

developing agronomic practices that address issues such as low yield and lodging in bere (Martin and Chang, 2007: 29). For a discussion of the nutritional properties of bere, see Theobald *et al.* (2006).

24 Evidence of the cultivation of the small oat as far back as the fifteenth century comes from archaeological sites in the UK (Scholten *et al.*, 2008: 1). Its use varies from animal (chiefly cattle) feed and human food to thatching and for making baskets and the seats of chairs. Scholten and colleagues note the continued dominance of the small oat, mixed with rye, on the machair in the late 1980s. They attribute its continued popularity up to this time to its limited need for fertilizer, its tolerance of soil deficiencies of manganese and copper, and the "cheapness of its home-produced seed" (Scholten *et al.*, 2008: 2).

25 *The Western Isles Local Biodiversity Action Plan* (2005) identifies the need to support seed production of local crop varieties for their inclusion in agro-ecological schemes (e.g. Comhairle nan Eilean Siar, 2005d; *LBAP, Cereal Fields and Margins Habitat Action Plan*), but thus far action has not been taken (Scholten, 2009: 17).

26 From analysis carried out at a number of sites in Lewis and Harris where there is evidence of woodlands, the Scottish Palaeoecological Archives Database gives a dating for birch of 7980–5030 BP and for Scots pine of 4870–3910 BP (Knott, 2003: 14).

27 The name, one of the few "purely" Gaelic place names – the majority reflect Norse as well as Gaelic influence – means "mouth" (Bun) of the "river" (Abhainn) that descends between two groups of hills (Eadarra) (Headland Archaeology and John Renshaw Architects, 2007: section 3.3.1, drawing on Lawson, 2006: 143).

28 In 1266, the Treaty of Perth marked the end of Norse lordship of the Outer Hebrides. Sovereignty was transferred to the Scottish Crown (Burgess, 2008: 34).

29 These activities were identified, among others, in the North Harris Interpretation Plan commissioned by the NHT in 2005 (Touchstone Heritage Management Consultants and Jo Scott, THMC and Scott, 2006). Building on earlier work carried out by the NHT and the John Muir Trust, the plan takes as its point of departure "the interplay between land and people, between the natural environment and the community that lives with it" (THMC and Scott, 2006: 6). The NHT required the consultants to carry out extensive community consultation.

30 The text is in English, the names on the map in Gaelic.

31 Much of the information concerning footpaths I owe to discussions with the NHT Land Manager, carried out on numerous occasions since 2003. In 2010, the NHT was successful in being awarded grants of £75 000 from the Scottish Rural Development Programme and £50 000 from Comhairle nan Eilean Siar for path improvement, the building of a small length of new path to connect the new bridge across the Abhainn Langadal to the Bogha

Glas–Mhiabhaig path, four car parking sites and interpretation panels at six sites. The NHT is itself contributing £35 000 to the project (*Caraidean*, Issue 9, August 2010: 2).

32 Since 2004, JMT volunteers have also been involved in beach clean-ups. West-facing beaches have, over time, accumulated substantial rubbish from places bordering the Atlantic from far afield, and efforts are now ongoing to keep the beaches clean. In the summer of 2010, the volunteers were also involved in the ongoing struggle to eradicate *Gunnera* ("giant rhubarb"), an invasive species introduced as an ornamental plant from South America 20 years ago. The plant is, according to Robin Read, the NHT ranger, together with climate change and the loss of habitat, "one of the biggest threats to biodiversity" (cited in *WHFP*, 8 October 2010: 3).

33 Tourism is an important income earner on Harris, the majority of visitors (94 per cent), totalling 30 857 in 2002, coming between the months of March and October (Western Isles Visitor Survey 2002, cited by NHT, 2007: 29). 60.9 per cent of these visitors are aged 45 years or more, and outdoor pursuits – hillwalking, bird watching, beach visits, photography and visiting historical sites – are listed as the most popular (NHT, 2007: 29).

34 The pamphlet was produced with the support of the John Muir Trust and SNH.

35 A site along the Chaolas Scalpay road is favoured as it is the closest to Tairbeart, and has a good wind regime (interview, Chair, North Harris Trading Company, 8 June 2010).

Spectral Lines, a small business now employing a staff of four, is often held up as an example of what can work on the island. Hugh MacPherson, the owner, moved his business from West Linton (20 miles south of Edinburgh) in the mid-1990s in order to reduce his costs. The site in Tairbeart provided the security he needed without exorbitant insurance costs. At that time, Spectral Lines focused on the design of microwave and radio frequency systems/communication systems that required high spectral purity, such as a navigation system for land mine detection. Since that time, the business has diversified and now includes electric and electronic work, information technology, metal working and woodworking.

The largest employer in North Harris is GSH, a branch of the parent company, which is based in Stoke on Trent. The branch in Tairbeart, employing 22 local people in 2011, is a data recovery centre and is also responsible for planning, administrative, audit and commercial contract services for the parent company. It also deals with human resource matters for GSH as a whole. Named after his father, George Scarr Hall, the company is owned by Ian Scarr Hall, whose purchase of Amhuinnsuidhe Castle Estate in 2003 facilitated the NHT's simultaneous purchase of the North Harris Estate (Chapter 2). The company's location here reflects Scarr Hall's ongoing commitment to the island's economic viability. While there is no impediment to the branch's location here, one employee commented that he saw the situation as "artificial" because, except for Scarr Hall, the jobs would be based in Stoke or carried out in homes where there were no overheads or benefits to be paid (interview, 31 March 2011).

36 As I have indicated earlier, the NHT also has links outwith the network of land owning community trusts. Through its long-standing relationship with the John Muir Trust – the John Muir Trust holds an invited directorship on the NHT board – the NHT is linked with broader discourses of "nature". Through its more recent connection with Carnegie UK's Community of Practice programme – whose aim is to facilitate the exchange of information on "community regeneration" with communities elsewhere in UK – the NHT is linked with the Centre for Alternative Technology, England; the Eden Project, Wales; and the Tipperary Institute, Ireland. Both the John Muir Trust and Carnegie were represented at the February 2011 meeting, as were Harris Development Limited and Highlands and Islands Enterprise.

6

Conclusion – Working Possibilities

Community Land Scotland/Fearann Coimhearsnacht na h'Alba

The story with which the book began has come full circle. Community Land Scotland/Fearann Coimhearsnacht na h-Alba was launched at the Inaugural General Meeting held in Inverness on 28 September 2010. The event was deep with historical resonance. Those present – the working group of three established at the meeting held a year before on Harris and representatives from many of the 17 community land owners who were now members – were all aware of the political significance of the new organization.[1] The land reform process remained stalled with a (minority) Scottish National Party (SNP) forming the government. No secure funding streams had been put in place. The Scottish Ministers had not yet decided on Pàirc Trust's high profile application under Part 3 of the Land Reform Act to purchase the common grazings of that estate and the associated interposed lease. And the West Harris Crofting Trust's protracted struggle with the government to purchase land in West Harris which was owned by the Scottish Ministers was still on people's minds.[2] The new organization was recognized as of both historical and contemporary significance. Said David Cameron, who together with Angela Williams and Lorne Macleod had been mandated to take forward the concerns from that earlier meeting on Harris,

Places of Possibility: Property, Nature and Community Land Ownership,
First Edition. A. Fiona D. Mackenzie.
© 2013 A. Fiona D. Mackenzie. Published 2013 by Blackwell Publishing Ltd.

Today we see a coming of age for community landownership in Scotland. Following a long history to achieve a right to buy, we want to work together to build on the benefits we are seeing from the early community pioneers. You can visit areas throughout Scotland which have been in community landownership for relatively few years and already there is evidence of new economic activity, you can feel a sense of re-invigoration and an increasing community confidence (cited in *WHFP*, 10 October 2010: 1).

Decisions about the form that the new organization would take, its structure of governance and remit were made following a period of extensive consultation with community land owners and the recommendations of the *Community Land Organisation Feasibility Study* (Bryan and Campbell, 2010).[3] Community Land Scotland was set up as a Company Limited by Guarantee with Charitable Status. Membership is open to all Scottish community organizations with "an active and stated interest in owning and managing land for the benefit of their community" (document circulated at the Inaugural General Meeting, 28 September 2010, Inverness). The board consists of up to 12 directors, of whom seven are elected from the membership and five may be co-opted. As identified at the Inaugural Meeting, Community Land Scotland's aims are:

1. To promote and represent the interests of community landowners across Scotland at all levels of local and national government and its agencies.
2. To promote the benefits of community landownership and to proactively encourage new landowners.
3. To collaborate with all other relevant membership and support organizations to ensure appropriate support services are provided to community landowners, in particular access to and uptake of existing services.
4. To facilitate networking and mutual support amongst community landowners.

While these aims are broad, Community Land Scotland has more recently specified in some detail the actions it proposes to take to fulfil its mandate. These include "engag[ing]" with the Scottish Government to ensure that communities' rights accorded under Parts 2 and 3 of the Land Reform Act 2003 are more easily accessed and creating a fund for community land purchase, which would draw from a variety of public and private sources (document circulated at the first Community Land Scotland Conference, Tairbeart, Isle of Harris, 29–30 March 2011).

The organization has one overriding concern, evident at the first Community Land Scotland Conference, held in Tairbeart, Isle of Harris, 29–30 March 2011, to ensure not only that the community land sector has greater visibility in the Scottish scene in its own right, but that the land reform process insofar as it pertains to community land ownership is taken up throughout Scotland (Photograph 6.1). Almost 500000 acres of land are now in community ownership, making this sector a major player in the national context. However, these acres are exclusively, as yet, in the Highlands and Islands. That 14 of the 17 members of the organization are located in the islands – with nine of these in the Outer Hebrides – signals even greater concentration.[4] The vast private estates of Ross-shire, Inverness-shire and Sutherland in the Highlands are, as yet, relatively but not totally untouched, as are those in the Borders. They remain as evidence that the land reform process still has a long way to go if the highly skewed distribution of land is to be addressed and if the process of land reform is not side-lined as relevant to the western Highlands and the Islands and nowhere else.

In the short time since its inauguration, Community Land Scotland has been active on the national as well as more local scenes. In the period prior to 1 May 2011, when a majority SNP government was elected, one of its most significant activities was to press the Scottish Government for a favourable decision on Pàirc's application under Part 3 of the Land

Photograph 6.1 The first Community Land Scotland Conference, 29–30 March 2011, Tairbeart.
Source: Used by kind permission of the photographer.

Reform Act. This outcome was achieved, although, as discussed in Chapter 4, community purchase of the estate is on hold, as the land owner has now placed the matter before the courts. In a highly visible move, the organization has taken the Rural Affairs Minister, Richard Lochhead, to task for the failure of a recently published document, *Speak Up for Rural Scotland* (Scottish Government, 2010), to identify the significance of community land ownership for addressing central concerns for rural Scotland: "the need for thriving communities, the need to enliven local democracy, the need for affordable housing and the need to take forward community renewable energy" (identified in Community Land Scotland press release, 8 November 2010). "[N]o link is made to the fact that community land ownership can facilitate and deliver all these changes", the press release continues. David Cameron, Chair, Community Land Scotland, emphasizes the point in the press release's conclusion:

> Community land ownership is a very significant success story and offers a completely different model not just in the Highlands and Islands but for the whole of Scotland. It should never have been written out of the script and we are determined to write it back in again.

In further initiatives, Community Land Scotland held meetings with the Scottish Affairs Committee and Members of the Scottish Parliament on the subjects of both community land ownership and the Crown Estate.[5] By inviting representatives of the parties contesting the May 2011 Scottish Government election in the Outer Hebrides to hustings at the conference of 29–30 March 2011 in Tairbeart, the organization contributed to raising the profile for land reform in the political scene. Meetings were also held with the Big Lottery, where the question of the use of funds from this source for the purchase of public land was raised, as well as the broader issue of financial support for community land purchase. The purchase of public land had proved to be a major sticking point in the case of the West Harris Crofting Trust's purchase of land owned by the Scottish Ministers, as I have already indicated. However, public land also includes land held by Forestry Scotland and the Ministry of Defence. Finally, lines of communication were established with groups where community land ownership is an option, the intent being to further the community-based land reform movement. As one example, Community Land Scotland was invited to a meeting on Scalpaigh, Harris, in early March 2011, where that community was deliberating whether to accept the offer of Fred Taylor, the owner, to hand over the island without cost – a move that has been contrasted with the actions of Barry Lomas regarding Pairc. Said David Cameron, speaking as Chair, Community Land Scotland, prior to that meeting,

> The reasoning and the ways in which the transfer is proposed is groundbreaking. It is extremely encouraging that Mr. Fred Taylor, the present owner, seeing the progress made by other landowning communities, is of the opinion that communities are in the best position to manage and develop their own land. The fact that he is willing to pass ownership to the community for absolutely nothing speaks volumes (cited in *WHFP*, 22 February 2011: 2).

The election of 1 May 2011 brought to power an SNP government in whose manifesto was a commitment to review the Land Reform Act 2003 and a proposal for a new Scottish Land Fund. Community Land Scotland has been quick in voicing its support. In September 2011, for example, in a press release (8 September 2011), accompanied by a detailed policy briefing, the organization called for the establishment of such a fund with a total value of £10 million to cover the four years of the parliament. Justifying their position, the press release states: "the evidence shows that when communities purchase the land on which they live and work, powerful new initiatives follow, halting decline, creating jobs and diversifying the local economy". As a further instance of their political engagement at the national level, Community Land Scotland has called for "the radical reform" of the Crown Estate (Community Land Scotland, press release, 15 September 2011, and accompanying policy briefing). In a political and policy context within Scotland, where dissatisfaction with the operation and management of the Crown Estate in Scotland has become ever more vocal, Community Land Scotland not only called for greater accountability to Scottish Ministers and the Scottish Parliament on the part of the Crown Estate, but additionally it has called for "locally accountable community land owners to take over the control of appropriate assets currently managed by the Crown Estate", whether these pertain to the land or adjoining shoreline and seabed (press release, 15 September 2011). This is indeed a bold move and has major implications, for example, with respect to leases for marine renewable energy projects (Community Land Scotland, The Crown Estate, policy briefing, 15 September 2011).

Community Land Scotland's rapidly growing visibility on the national scene, I would suggest, vindicates the decision taken, subsequent to the *Feasibility Study* (Bryan and Campbell, 2010), to establish an organization whose exclusive focus was community land ownership. There had been some discussion about becoming a member of an overarching organization, Scottish Community Alliance, whose remit is more general. However, for David Cameron, it was necessary to stand "on [our] own two feet as a sector rather than diluting the cause by joining with another group" (interview, 2 October 2009). Having said this, Community Land Scotland's actions make clear that it places considerable importance in liaising with other community-based groups and in working with them when it is mutually advantageous to do so. Recent communication with the Community Woodland Association provides one example. A grant of £60 000 over three years from Highlands and Islands Enterprise in March 2011, together with one of £10 000 from Awards for All, to cover mainly administrative costs, including employing a part-time Networking Co-ordinator and a part-time Public Affairs and Representation Manager, has provided Community Land Scotland with the initial capital necessary to extend and deepen its work of furthering the process of community-based land reform in Scotland (Highlands and Islands Enterprise press release, 22 March 2011).

Property, nature and community land ownership

Possibility is not a luxury; it is as crucial as bread
(Butler, 2004: 29).

My hope in writing this book has been to make visible a narrative that searches for an alternative to a process of neoliberal globalization whose frequently devastating effects have been the subject of substantial critical social science analysis. My concern has been to contribute to a call to redress an imbalance that has resulted from this research (Castree, 2008a: 143) – namely that significantly more energy has been spent on documenting the process of dispossession than in tracing resistance or alternatives to it. I have sought to address the question "what is already being done" as a counter-move to global processes of enclosure and privatization rather than the question "what is to be done", thereby, as J. K. Gibson-Graham and Gerda Roelvink (2010: 331) suggest with respect to a community economy, "contributing to the credibility and strengthening" of a different way of seeing.

By engaging in research into the specificities of a particular place, the Outer Hebrides, at a particular point in time, and linking the happenings here with broader debates about property and nature, my objective has been to tease out the meanings of community land ownership for this counternarrative, to trace the ways in which a commoning of the land challenges the hegemony of neoliberal discourse. Community Land Scotland, representing the interests of communities that have struggled, sometimes against great odds, to re-claim a land that has been the subject of enclosure and privatization for hundreds of years, provides a significant marker along this route. Its transformative possibilities are, as of yet, barely identified. They include extending the geographical reach of community land ownership beyond the western Highlands and Islands to Scotland as a whole and supporting activities that contribute to the building or strengthening of more socially and environmentally just communities and economies than was previously the case. They are linked, discursively, to a global movement that, however "hydra-like" (Castree *et al.*, 2010: 5), challenges the neoliberal agenda and conjures in its place more socially, environmentally and economically generous "postneoliberalisms" (see Peck *et al.*, 2010: 110).

The significance of Community Land Scotland and its constituent members for this broader, very plural, movement as well as for Scotland itself lies in its members' disruption of seemingly unassailable property norms. As recalled through the stories told in this book, a commoning of the land through community ownership interrupts an ontology that produces power/knowledge in ways that "work together to establish a set of subtle and explicit criteria for thinking the world" (Butler, 2004: 27). Community land ownership interrupts the givenness of private property and thereby the process of privatization at the centre of neoliberalization. It calls into question an optic confined by the binary public/private, the so-called ownership model of property, premised on a bundle of rights that are claimed as exclusive and absolute, fixed and formulaic, however porous and permeable in practice they may turn out to be (Blomley, 2004).

With a commoning of the land, the politics of property is opened up to new material and metaphorical configurations. To follow Butler's (2004: 37–38) line of argument, which she traces with respect to the categories human, gender and human rights, this new politics depends on using the language of a common right "to assert an entitlement" to the land while at the same time "subject[ing]" the category common property to continuous "critical scrutiny". Property held in common is, like the categories on which Butler focuses, always "in process", and "thus we do not yet know and cannot ever definitively know in what [common property] finally consists" (Butler, 2004: 37). It may be, again to draw from Butler (2004: 231), that some norms will be "useful" for this new politics, "but they will be norms that no one will own, norms that will have to work not through normalization...but through becoming collective sites of continuous political labor".

Community land owning trusts provide the political space for such collective labour. It is here that property norms evident in the ownership model are interrupted and new ways of seeing the land come into being. To use Blomley's (2008: 322) words, it is here that a "space of hope and potentiality is prised open". Or, to borrow from his citation of an activist in Vancouver's Downtown Eastside, what was once a "place of impossibilities" (Blomley, 2004: 62) becomes instead a "place of possibilities" (Mackenzie, 2010b: 339). Initially, it is through the trusts' structures of governance that a collective claim to the land is struggled over, that new meanings of the age-old right to land captured in the metaphor of dùthchas are created. In this respect, dùthchas has proved to be a mobile signifier. It is both an inherited right which, at the time of the Clearances, informed visible resistance to processes of enclosure and privatization, and an evolving right, which now legitimates a claim of community, not all of whose members had pre-existing rights to the land (see Nash, 2002). As deliberations over the community's purchase of the North Harris Estate made evident, a claim to the land was not restricted to those who could trace genealogical depth on the island or who, nominally at least, worked the land designated as under crofting tenure. As I discussed in Chapter 2, dùthchas has provided the discursive means to translate an historically resonant collective right to the land into a contemporary place-based right of social justice. All residents on the estate have the right to become members of the owning collectivity. To recall Blomley's (2008: 318) argument about "the property of the poor" in a context of gentrification in Vancouver, a commoning of the land may be a material struggle, but it is also about claiming "a moral and political commons" – in Scots, a commonweal.

That the process of commoning the land where crofting tenure prevails may contain the seeds of contradiction is evident. I have drawn attention to the provisions of the Crofting Reform Act 1976, subsequently included in the consolidating legislation of the Crofters Act 1993, which allow for individual croft ownership. By underpinning the marketization of crofts, the Act of 1993 has worked against the intent of the Land Reform Act 2003. Particularly where land owned by communities lies solely within crofting tenure (as, for example, is the case in Galson and West Harris, but not in North Harris), the material base of collective ownership is potentially under threat (MacAskill, 2004). The extent to which the Crofting Reform Act 2010 provides the legal means of effectively redressing this situation insofar as it concerns the marketization of crofts and the speculative sale of croft land for housing sites remains to be seen.

Subsequent to community purchase, the meanings of property held in common are constantly reworked at monthly or bi-monthly board meetings, the annual general meetings of both the land owning trust and any wholly owned trading companies, during public meetings held to discuss specific proposals such as a wind farm or a national park, meetings with crofters to ascertain the availability of land designated as common grazings for affordable housing, or during such activities as the planting of native woodland, the restoration of a network of paths, an archaeological dig and the production of local food. It is here – in more formal as well as more everyday encounters – that property held in common is constantly a-doing, continually re-conjured, literally and figuratively, in the interest of a collectivity rather than in the interest of a private individual or a corporation.

This collective and continuous a-doing of property is bound up with the becoming of a collective subjectivity. Where a common right to the land is claimed, in other words, community and property are co-constituted. The practices of governance of the community trust and the bye-laws that stipulate the rules of membership are central to this. Together, they create a discursive space where a collective identity is re-forged and categories of belonging become less coherent and more "transformable" (see Butler, 2004: 216). All adults who live on the North Harris Estate may choose to become members of the land owning trust – the membership fee is £1 – and thus owners in common of the land, that which is under crofting tenure and, in the case of the NHT, that which is not. It is they, collectively, who now exercise responsibility towards the land in its entirety. The Assynt Crofters' Trust, where crofters alone hold ownership rights, a matter which for some remains a source of dispute, is an exception to this general practice. Elsewhere, for crofters, the process of resubjectivation is complicated by the co-working of rights under crofting tenure and those crafted through practices associated with the provisions of the Land Reform Act, evident in the legal requirements of a company limited by guarantee. As I have discussed in Chapter 2, with collective ownership of the land, a crofter negotiates the subject position of tenant as defined by crofting law, with use rights to inbye land and common grazings, at the same time as working the potentially contradictory subject position of owner of a community owned estate, a position shared with non-crofters. That this unsettling of the boundaries between tenant and owner of the land can lead to dispute is demonstrated by the case of Stòras Uibhist and the restoration of the Askernish golf course.

At the same time that land owning community trusts become the sites of collective labour where property norms are "undone", so too do they become the sites where, through collective labour, the norming of nature through the nature/culture divide is unsettled. Whether it concerns conservation, on the one hand, or complicating processes of commodification, on the other, the disturbance of property norms opens nature to the possibility of more socially just and sustainable significations than was previously the case. With respect to conservation, the NHT's initiatives in planting native woodland, in signing a management agreement with SNH, and in proposing a national park for the whole of the Isle of Harris, "undo" nature – and the wild – as inert political categories fixed through the Great Divide. Through each, the wild is recalled, not as an essentialist or original category bound to conservationist imperative or class interest, but as one further way through which the land is or could be worked. The initiatives expose nature and the wild as contingent categories whose meanings and materiality are produced through the interworking of the human and the non-human in the discursive spaces opened up by a land owning community trust. They point towards a politics that traces connections and complexities, a politics of social nature that, to use Braun's (2002: 13, emphasis his) words again, positions people "*in* nature", not external to it.

From this position, the NHT, Galson and Stòras Uibhist, as examples, engage in a politics of nature that troubles the norms of commodification through the capturing of the wind or water for the generation of renewable energy. Whereas privatization of the wind and the construction of large-scale wind farms through corporate efforts or those of private individuals – where they frustrate the aspirations of crofting communities to own the land on which they reside – are unequivocally part and parcel of a politics that proceeds through a process of "accumulation by dispossession" (Harvey, 2003), the community trusts' politics of small-scale wind farms or hydro installations counters this direction. David Cameron's words, as Chair of the North Harris Trading Company, cited in Chapter 4, make this clear. Community control of a wind farm was, he stated, essential and land ownership was "critical" for turning things around (interview, 20 June 2004).

In this politics, norms of commodification are disrupted. Surplus is no longer distributed to a company's shareholders or directed into a private individual's bank accounts. Instead, it is appropriated by a community land owning trust and distributed according to its democratically debated priorities. Thus, where there is congruence between claims to a new energy commons and to a new property commons, the wind and water become the means through which collective rights to land are affirmed, not the means through which people are dispossessed.

Just as people rework their subject positions through the ongoing realization of rights in common to the land, so does the new contingent politics of nature provide the discursive space for reworking individual and collective subjectivities. Community "becomes" through debate – about national park status, wind turbines, small-scale hydro installations, house insulation; it also becomes through action (see Macdonald, 1997; Kohn, 2002). The work parties that plant native woodland, that repair footpaths or that collect the accumulated debris brought in by winter storms from the beaches, and the planting of fruit and vegetables in polytunnels by youth now involved in Crofting Connections, all provide routes. These debates and activities replace sheep as the "glue" that previously held crofting communities together.[6] Or, as one research participant said to me, it was now wind turbines rather than crofting that brought people together. Together with the more explicit reworking of property rights identified earlier, they expand the discursive repertoire through which people reposition themselves collectively. Crofting is no longer the only signifier of belonging but one of a number of possible subject positions through which people negotiate their collective identity and re-create a place that is continuously "becoming".

Despite the short time period during which community land ownership has been practised in the Highlands and Islands, the evidence suggests that the process of collective resubjectivation is tied to the beginnings of a community economy. While this may take the dramatic form of Stòras Uibhist's ambitious proposals for the rehabilitation of the Lochboisdale harbour area, it more commonly includes such ventures as small-scale wind farms and hydro installations, a recycling facility, the planting of native woodland, the provision of facilities for light industry, the extension of horticultural production and the restoration of footpath networks, as I have discussed. Together, these and other initiatives are making visible an "econo-sociality" (Gibson-Graham and Roelvink, 2010: 330) that furthers economic diversity and contributes to the achievement of more locally directed sustainable futures. As is evident in the NHT's support for affordable housing, "the crucial interdependence of economic and 'noneconomic' activities" (Gibson-Graham, 2006: 95) has been recognized.

It is on these grounds – of a reworking of property rights and nature in the collective interest – that the case can be made that this process of collective resubjectivation where land is in community ownership is not about the cooption of community for a neoliberal agenda but is instead about a refusal of the terms of that agenda (see Katz, 2005: 631). Where the land is a-commoning, the becoming of a community subjectivity reverses processes of enclosure and privatization; it does not contribute to their extension. It relies on a politics that keeps open to continuous re-signification the meanings of property and nature, a politics that, to borrow once again from Butler (2004: 222, replacing the term "human" in the original), provides the discursive space to think and act "critically – and ethically – about the consequential ways that [property and nature are] being produced, reproduced, deproduced". To follow Butler's analysis of the political possibilities created through the disruption of norms, it is a politics that exposes interruptions in the ways in which property and nature are given to be seen, opening up the possibilities of community land ownership to new, more sustainable, socially and environmentally just, imaginaries than those possible through processes and practices of private land tenure.

Evidence continues to accumulate that community land ownership "works", that processes of dispossession that had previously defined the Outer Hebrides are being countered. What had previously been considered impossible is now in the process of becoming possible through a commoning of the land. Primarily through a case study of the North Harris Trust, I have explored how community land ownership, connected as it undoubtedly is to a deep sense of historical justice and a collective right of belonging to the land, is leading to socially, culturally, environmentally and economically "resilient" (see Skerratt, 2011) communities. The actions of the communities themselves and now of Community Land Scotland contribute to the growing visibility of this counterhegemonic narrative, discursively displacing that informed by neoliberal doctrine. They suggest that there is indeed a different, more socially just, and sustainable means of constituting the way forwards.

Notes

1 In September 2010, membership consisted of the following: Isle of Eigg Heritage Trust, Isle of Gigha Heritage Trust, Knoydart Foundation, Borve and Annishadder Township, The Stornoway Trust, the Assynt Foundation, Urras Oighreachd Ghabhsainn, Bhaltos Community Trust, Stòras Uibhist, North West Mull Community Woodland Company, West Harris Crofting Trust, Isle of Rum Community Trust, North Harris Trust, the Pàirc Trust, Mangistera Trust, Urras Sgire Oighreachd Bharabhais and Aigas Community Forest.
2 The land in question comprised three estates, which the Department of Agriculture for Scotland had acquired in the aftermath of World War I – land *Fit for Heroes* (Leneman, 1989) – Scaristavore in 1929, Borve in 1933 and Losgaintir in 1943 (CIB Services, 2008b: 8). In total, the estates comprise 6604 acres (2674 ha), all under crofting tenure. A ballot of the 126 adults

resident on the estates indicated "overwhelming" support for a community buyout of the land: 94 per cent of those eligible cast a vote, 77 per cent of those voting expressing support (*WHFP*, 10 October 2008: 2). As government owned property, the transfer of land to a community body fell under the provisions of the Transfer of Crofting Estates (Scotland) Act 1997, which does not, *inter alia*, require a community to pay for the land. Nevertheless, for reasons that the local Member of the Scottish Parliament Alasdair Allan explained as "strict accounting rules which had to be adhered to and the legal issues involved" (*Stornoway Gazette*, 21 May 2009: 3), the West Harris Crofting Trust (WHCT) faced a government that insisted that the land be paid for at full market value, £59 000. However, whereas other communities had been able to access funds from public sources to cover almost completely the costs of purchase, the WHCT found that it was not allowed access to these sources as the land the community sought to purchase was already "publicly" owned (interview, Chair, WHCT, 18 December 2008). The government's position seemed all the more problematic as, around the same time, with the enthusiastic endorsement of Michael Russell, the Scottish Government's Environment Minister, SNH had offered the land and buildings of the island of Rum to its residents effectively without cost (Macleod and Wade, 2009: 21). In the event, the WHCT finally secured funding from HIE and Comhairle nan Eilean Siar and a loan from Tighean Innse Galle, the local housing agency. The land came into community ownership in January 2010.

3   With a return of 70 per cent, the survey carried out for the *Feasibility Study*, funded by Highlands and Islands Enterprise, indicated unanimous support for an organization for community land owners.

4   Of the nine in the Outer Hebrides, two – Pairc and Mangistera – do not yet own land. It is of note that the Assynt Crofters' Trust, despite having accepted an invitation to attend the Inaugural Meeting in Inverness, has not as yet joined Community Land Scotland.

5   Regarding the continuing antagonism between local communities and the Crown Estate, see Wightman (2010); also, MacAskill (2006).

6   The words were those of a participant in the Crofting Inquiry chaired by Mark Shucksmith and cited by him at the Scottish Crofting Federation Conference, Dingwall, 6 September 2007.

References

Abromowitz, D.M. (2000) An essay on community land trusts: toward permanently affordable housing, in *Property and Values: Alternatives to Public and Private Ownership* (eds C. Geisler and G. Daneker), Island, Washington, DC, pp. 213–232.

Adkin, L. (1992) Counter-hegemony and environmental politics in Canada, in *Organizing Dissent. Contemporary Social Movements in Theory and Practice* (ed. W. K. Carroll), Garamond, Toronto, pp. 135–156.

Aquaterra (no date) *Creating Connections. A Case for Renewables Investment and Grid Connection in Orkney, Shetland and the Outer Hebrides*, pamphlet. Contact details: Shetland, Aaron Priest (aaron.priest@sic.shetland.gov.uk); Orkney, Jeremy Baster (jeremy.baster@orkney.gov.uk); Western Isles, Derek McKim (dmckim@cne-siar.gov.uk).

Armit, I. (1996) *The Archaeology of Skye and the Western Isles*, Edinburgh University Press, Edinburgh.

Ashmore, P. (2002) *Calanais: The Standing Stones*, Historic Scotland, Edinburgh.

Bakker, K. (2007a) The "commons" versus the "commodity": alter-globalization, anti-privatization, and the human right to water in the global south. *Antipode*, 39 (3), 430–455.

Bakker, K. (2007b) Neoliberalising nature? Market environmentalism in water supply in England and Wales, in *Neoliberal Environments. False Promises and Unnatural Consequences* (eds N. Heynen, J. McCarthy, S. Prudham and P. Robbins), Routledge, Abingdon, pp. 101–113.

Barbor-Might, D. (2008) Food from the edge. *The Scotsman*, 30 April: 3.

Barker, A. and Stockdale, A. (2008) Out of the wilderness? Achieving sustainable development within Scottish national parks. *Journal of Environmental Management*, 88, 181–193.

Barton, H. (1996) The Isle of Harris Superquarry: concepts of the environment and sustainability. *Environmental Values*, 5, 97–122.

Beinn Mhor Power (BMP) (2007) *Feiriosbhal*, proposal submitted to Comhairle nan Eilean Siar, Stornoway.

Bell, E. (2007) Crofters appeal to Scottish Land Court. *Stornoway Gazette*, 12 April, 1.

Blomley, N. (2002) Mud for the land. *Public Culture*, 14 (3), 557–582.

Blomley, N. (2003) Law, property, and the geography of violence: the frontier, the survey and the grid. *Annals of the Association of American Geographers*, 93, 121–141.

Blomley, N. (2004) *Unsettling the City. Urban Land and the Politics of Property*, Routledge, New York.

Blomley, N. (2005) Remember property? *Progress in Human Geography*, 29 (2), 125–127.

Blomley, N. (2008) Enclosure, common right and the property of the poor. *Social and Legal Studies*, 17 (3), 311–331.

Braun, B. (2002) *The Intemperate Rainforest*, University of Minnesota Press, Minneapolis, MN.

Brien, R. (1989) *The Shaping of Scotland*, Aberdeen University Press, Aberdeen.

Bright, J.A., Langston, R.H.W., Bullman, R., Evans, R.J., Gardner, S. and Pearce-Higgins, J., Wilson, E. (2006) Bird sensitivity map to provide locational guidance for onshore wind farms in Scotland. *RSPB Research Report 20*.

Bright, J.A., Langston, R.H.W., Bullman, R., Evans, R.J., Gardner, S. and Pearce-Higgins, J. (2008) Map of bird sensitivities to wind farms in Scotland: a tool to aid planning and conservation. *Biological Conservation*, 141, 2342–2356.

Brown, K.M. (2007) Understanding the materialities and moralities of property: reworking collective claims to land. *Transactions of the Institute of British Geographers NS*, 32, 507–522.

Bryan, A. and Campbell, L. (2010) *Community Land Organisation Feasibility Study*, Highlands and Islands Enterprise, Inverness.

Bryden, D., Westbrook, S., Taylor, B. and Bell, C. (2008) Isle of Harris National Park: feasibility study. *Final Report. Summary, Main and Appendices for the Isle of Harris National Park Study Group*.

Bryden, J. (2007) Sustainable rural communities in crofting areas. *Paper Commissioned by the Committee of Inquiry on Crofting*, www.croftinginquiry.org (last accessed 28 March 2012).

Bryden, J. and Geisler, C. (2007) Community-based land reform: lessons from Scotland. *Land Use Policy*, 24 (1), 24–34.

Buchanan, J. (1996) *The Lewis Land Struggle. Na Gaisgich*, Acair, Stornoway.

Burgess, C. (2008) *Ancient Lewis and Harris. Exploring the Archaeology of the Outer Hebrides*, Comhairle nan Eilean Siar, Stornoway.

Busby, N. and Macleod, C. (2010) Rural identity in the twenty-first century: a community of crofters or crofting communities. *Journal of Law and Society*, 37 (4), 592–619.

Butler, J. (2004) *Undoing Gender*, Routledge, New York.

Callander, R. (1998) *How Scotland is Owned*, Canongate, Edinburgh.

Cameron, D. (2006) Proposed wind turbines at Monan, Isle of Harris. Letter to Mairi MacKinnon, Department of Sustainable Communities, Comhairle nan Eilean Siar, 5 November.

Cameron, D. (2007) Proposed wind turbines at Monan, Isle of Harris. Letter to Mairi MacKinnon, Department of Sustainable Communities, Comhairle nan Eilean Siar, 22 February.

Cameron, D. (2009) "Old and new" working for each other on community land, Letter to the Editor, *West Highland Free Press*, 9 October, 15.

Cameron, E.A. (2001) "Unfinished business": the Land Question and the Scottish Parliament. *Contemporary British History*, 15 (1), 83–114.

Campbell, A. (2009) No quick fixes for shortage of affordable homes. *West Highland Free Press*, 22 May, 15.

Campbell, G. (2007) Letter to the Editor. *West Highland Free Press*, 2 February, 11.

Caraidean (2010) issue 9, August, 4.

Caraidean (2010) issue 9, August, 2.

Carey Miller, D.L. and Combe, M.M. (2006) The boundaries of property rights in Scots Law. *Electronic Journal of Comparative Law*, 10 (3), 3–24.

Carver, S. (2009) Mapping wildness in the Cairngorm National Park. *Wild Land News*, Spring, 10–16.

Carver, S., Fritz, S., Comber, L., McMorran, R. and Washtell, J. (2008) *Wildness Study in the Cairngorms National Park*, commissioned by the Cairngorms National Park Authority and Scottish Natural Heritage.

Casid, J.H. (2005) *Sowing Empire: Landscape and Colonization*, University of Minnesota Press, Minneapolis, MN.

Castree, N. (2003) Commodifying what nature? *Progress in Human Geography*, 27(3), 273–297.

Castree, N. (2004) Differential geographies: place, indigenous rights and "local" resources. *Political Geography*, 23, 133–167.

Castree, N. (2008a) Neoliberalizing nature: the logics of deregulation and reregulation. *Environment and Planning A*, 40, 131–152.

Castree, N. (2008b) Neoliberalizing nature: processes, effects, and evaluations. *Environment and Planning A*, 40, 153–173.

Castree, N. and Braun, B. (1998) The construction of nature and the nature of construction, in *Remaking Reality. Nature at the Millennium* (eds B. Braun and N. Castree), Routledge, London, pp. 3–42.

Castree, N., Chatterton, P., Heynen, N., Larner, W. and Wright, M.W. (2010) Introduction: the point is to change it, in *The Point Is To Change It* (eds N. Castree, P. Chatterton, N. Heynen, W. Larner and M.W. Wright), Wiley-Blackwell, Chichester, pp. 1–9.

CIB Services (2008a) *North Harris Trust Business Plan 2007–2018*, commissioned by the North Harris Trust.

CIB Services (2008b) *West Harris Community Buyout. Feasibility Study*, commissioned by the West Harris Crofting Trust.

Climate Change (Scotland) Act 2009 (2009) asp 12, Stationary Office, Edinburgh.

Colls, K., Hunter, J. and Colls, C.S. (2010) *Harris Community Archaeology Project 2010: Activity Report*. Birmingham Archaeology, University of Birmingham.

Combe, M.M. (2006) Parts 2 and 3 of the Land Reform (Scotland) Act 2003: a definitive answer to the Scottish Land Question? *Juridical Review*, 3, 195–227.

Comhairle nan Eilean Siar (CnES) (2003) *Western Isles Structure Plan*, CnES, Stornoway.

Comhairle nan Eilean Siar (CnES) (2004) *Ar Nàdar. Frèam airson Bith-ionadachd sna h-Eileanan Siar/Our Nature – A Framework for Biodiversity Action in the Western Isles*, CnES, Stornoway.

Comhairle nan Eilean Siar (CnES) (2005a) *Lewis Windfarm Proposal by Lewis Wind Power*, considerations and submissions by CnES, Stornoway.

Comhairle nan Eilean Siar (CnES) (2005b) *Muaitheabhal Windfarm Proposal by Beinn Mhor Power*, considerations and submissions by CnES, Stornoway.

Comhairle nan Eilean Siar (CnES) (2005c) Additional papers provided to the Environmental Services Committee, 28 June 2005, Stornoway.

Comhairle nan Eilean Siar (CnES) (2005d) *Cereal Fields and Margins Habitat Action Plan/Plan-gníomha árain is Achaidhean is Oirean Arbhair*, Western Isles Local Biodiversity Action Plan/Plan-gníomha Bithiomadachd Ionadail nan Eilean Siar, Stornoway.

Comhairle nan Eilean Siar (CnES) (2007a) Papers provided to the Environmental Services Committee, 8 February 2007. Re Addendum to Lewis Wind Power Ltd Wind Farm proposals, Stornoway.

Comhairle nan Eilean Siar (CnES) (2007b) Papers provided by Director for Sustainable Communities to the Environmental Services Committee, Erect 3 wind turbines and associated transformers and access track, Monan, North Harris, 22 March, 107–149.

Comhairle nan Eilean Siar (CnES) (2008a) *Feiriosbhal Wind Farm – Report by Head of Development Services*, February.

Comhairle nan Eilean Siar (CnES) (2008b) Local food conference. *Report by Chief Executive, Sustainable Development Committee*, 22 October.

Comhairle nan Eilean Siar (CnES) (2010a) Proposed Isle of Harris National Park, *Report by Director of Development, Sustainable Development Committee*, 21 April.

Comhairle nan Eilean Siar (CnES) (2010b) Proposed Isle of Harris National Park, *Report by Director of Development, Sustainable Development Committee*, 27 October.

Committee of Inquiry on Crofting (2008) *Final Report* (Chair M. Shucksmith), Edinburgh.

Community Energy Scotland (CES) (2008) *Community Energy News*.

Community Energy Scotland (CES) (2009) *Annual Review 2008/2009*.

Community Energy Scotland (CES) (2010) *Community Energy News*, Spring/Summer.

Connolly, W. (1999) *Why I Am Not a Secularist*, University of Minnesota Press, Minneapolis, MN.

Cosgrove, P.J. and Farquhar, J.E. (1999) Distribution and conservation status of the freshwater pearl mussel *Margaritifera margaritifera* in Lewis and Harris. *ROAME Reference F99LC02. A Report to SNH. Contract HT/ LCO2/99/00/36.*

Cosgrove, P., Hastie, L. and Young, M. (2000) Freshwater pearl mussels in peril. *British Wildlife*, 11 (5), 340–347.

Countryside Commission for Scotland (1990) *Management of the Mountain Areas of Scotland*, Battleby, Perth.

Cowen, M.P. and Shenton, R.W. (1996) *Doctrines of Development*, Routledge, London.

Crofter (2009) November, 9.

Crofters Commission (2011) http://www.crofterscommission.org.uk/What-is-Crofting.asp (last accessed 28 March 2012).

Crofting Reform (Scotland) Act 2010 (2010) asp 14, Stationery Office, Edinburgh.

Cronon, W. (1995) The trouble with wilderness; or, getting back to the wrong nature, in *Uncommon Ground: Toward Reinventing Nature* (ed. W. Cronon), Norton, New York.

Cruikshank, J. (2005) *Do Glaciers Listen? Local Knowledge, Colonial Encounters, and Social Imagination*, UBC Press, Vancouver, and University of Washington Press, Seattle, WA.

Cunningham, P. (2000) A new woodland walk in Harris. *Stornoway Gazette*, 16 November, 5.

Currie, B. (2003) The wind of change blows through North Harris. *West Highland Free Press*, 28 March, 7.

Curtis, R. and Curtis, M. (2008) *Callanish. Stones, Moon and Sacred Landscape. Windfarm Submission*, Author, Callanish, UK).

Dè Tha Dol? (2006) 17 February, 9.

Desmarais, A.A. (2008) The power of peasants: reflections on the meanings of La Vía Campesina. *Journal of Rural Studies*, 24, 138–149.

Devine, T. (1994) *Clanship to Crofters' War: The Social Transformation of the Scottish Highlands*, Manchester University Press, Manchester.

Dewar, D. (1998) *Land Reform for the 21st Century*, The Fifth John McEwen Memorial Lecture on Land Tenure in Scotland, A.K. Bell Library, Perth.

Dressler, C. 1998. *Eigg. The Story of an Island*, Polygon, Edinburgh.

Dreyfus, H.L. and Rabinow, P. (1983) *Michel Foucault: Beyond Structuralism and Hermeneutics*, University of Chicago Press, Chicago, IL.

DuPuis, E.M. and Goodman, D. (2005) "Should we go home to eat?": toward a reflexive politics of localism. *Journal of Rural Studies*, 21, 359–371.

Dwelly, E. (1994) *Faclair Gaidhlig Gu Beurla Le Dealban/Illustrated Gaelic to English Dictionary*, Gairm, Glasgow.

Edwards, K.J., Whittington, G. and Ritchie, W. (2005) The possible role of humans in the early stages of machair evolution: palaeoenvironmental investigations in the Outer Hebrides, Scotland. *Journal of Archaeological Science*, 32, 435–449.

Edwards, T. (2010) *SPICe Briefing on Crofting Reform (Scotland) Bill 2010.*
Available from The Scottish Parliament/Parlamaid na h-Alba.

Escobar, A. (1995) Imagining a post-development era, in *Power of Development*
(ed. J. Crush), Routledge, London, pp. 211–228.

Escobar, A. (1996) Constructing Nature. Elements for a poststructural political
ecology, in *Liberation Ecologies* (eds R. Peet and M. Watts), Routledge,
London, pp. 46–68.

Faber Maunsell (2008) *The Western Isles Hydropower Feasibility Study*,
Highlands and Islands Energy Company, Inverness.

Ferguson, M. and Forster, J.A. (2005) Establishing the Cairngorms National
Park: lessons learned and challenges ahead, in *Mountains of Northern Europe.
Conservation, Management, People and Nature*, TSO Scotland, Edinburgh,
pp. 275–289.

Ferguson, R. (2008) Self sufficient islands. *Stornoway Gazette*, 17 January, 8.

Fitzsimmons, M. (2004) Engaging ecologies, in *Envisioning Human Geographies*
(eds P. Cloke, P. Crang and M. Goodwin), Arnold, London, pp. 30–47.

Flanagan, T. and Alcantara, C. (2004) Individual property rights on Canadian
Indian reserves. *Queen's Law Journal*, 29 (2), 489–532.

Fletcher, T. (2002) Reforesting Harris. *CLAN Bulletin*, p. 10.

Forestry Commission Scotland (2000) *Forests for Scotland*, Forestry Commission
Scotland, Edinburgh.

Forestry Commission Scotland (2004) *Western Isles Locational Premium*, Forestry
Commission Scotland, Edinburgh.

Forestry Commission Scotland (2005) *Scottish Forestry Grants Scheme*, Forestry
Commission Scotland, Edinburgh.

Forestry Commission Scotland (2006) *The Scottish Forestry Strategy*, Forestry
Commission Scotland, Edinburgh.

Foucault, M. (1979) *Discipline and Punish: The Birth of the Prison*, Vintage,
New York.

Foucault, M. (1985) *The History of Sexuality: An Introduction*, Vol. 1, Vintage,
New York.

Foucault, M. (1990) *The History of Sexuality: The Use of Pleasure*, Vol. 2, Vintage,
New York.

Foucault, M. (2007) What is critique? in *The Politics of Truth* (eds S. Lotringer
and L. Hochroch), Semiotext(e), New York, pp. 41–82.

Fraser, M. (2004) Protesters torch a turbine effigy. *Stornoway Gazette*,
16 December, 1.

Fraser, S. (2010) Improving the legislation? in *Land Reform. Rights to Buy –
Where To from Here?* conference report, 2010, Highland Council Chambers,
Inverness, pp. 22–24.

Gerrard, N. (2010) Achievements in community land ownership in the
Highlands and Islands, in *Land Reform. Rights to Buy – Where To from
Here?* conference report, 2010, Highland Council Chambers, Inverness,
pp. 16–17.

Gibson-Graham, J.K. (2003) An ethics of the local. *Rethinking Marxism*, 15, 49–74.

Gibson-Graham, J.K. (2006) *A Postcapitalist Politics*, University of Minnesota Press, Minneapolis, MN.

Gibson-Graham, J.K. and Roelvink, G. (2010) An economic ethics for the Anthropocene, in *The Point Is To Change It* (eds N. Castree, P. Chatterton, N. Heynen, W. Larner and M. Wright), Wiley-Blackwell, Chichester, pp. 320–346.

Ginn, F. (2008) Extension, subversion, containment: eco-nationalism and (post) colonial Nature in Aotearoa New Zealand. *Transactions of the Institute of British Geographers NS*, 33, 335–353.

Gordon, W.M. (1999) *Scottish Land Law*, Green, Edinburgh.

Graeme Scott and Co. (2002) *North Harris Estate feasibility study*. Final Report for the Steering Group.

Graeme Scott and Co. (2004) *Loch Seaforth Estate feasibility study*. Final Report for the Steering Group.

Grant, I.F. (1961) *Highland Folk Ways*, Routledge and Kegan Paul, London.

Gregory, D. (2000) Edward Said's imaginative geographies, in *Thinking Space* (eds M. Crang and N. Thrift), Routledge, London, pp. 302–348.

Gregory, D. (2004) *The Colonial Present*, Blackwell, Malden.

Gregory, R.A., Murphy, E.M., Church, M.J., Edwards, K.J., Guttmann, E.B. and Simpson, D.D.A. (2005) Archaeological evidence for the first Mesolithic occupation of the Western Isles of Scotland. *The Holocene*, 15 (7), 944–950.

Halcrow Group Ltd. (2009) *Economic and Community Benefit Study*. Final Report for the Scottish Government.

HallAitken (2007) *Outer Hebrides Migration Study. Final Report*, commissioned by the Western Isles Council/Comhairle nan Eilean Siar.

Hamilton, J. (2010) Discovering the past to understand the future. *The Crofter*, 86, 11.

Hardin G. (1968) The tragedy of the commons. *Science*, 162, 1243–1248.

Harris National Park Study Group (2010) *Minutes of Meeting*, 24 March.

Hart, G. (2004) Geography and development: critical ethnographies. *Progress in Human Geography*, 28, 91–100.

Harvey, D. (2003) *The New Imperialism*, Oxford University Press, Oxford.

Harvey, D. (2005) *A Brief History of Neoliberalism*, Oxford University Press, Oxford.

Hastie, L. (2004) Freshwater pearl mussel survey of the North Harris (---------River) candidate Special Area of Conservation. *Confidential Scottish Natural Heritage Research Report F02PA04c*.

Hastie, L.C. and Young, M.R. (2003) *Conservation of the Freshwater Pearl Mussel 2. Relationship with Salmonids, Conserving Natura 2000 Rivers Conservation Techniques Series 3*, English Nature, Peterborough.

Haworth Conservation Ltd. (2006) *Monan Windfarm. Ornithological Survey and Assessment*, for the North Harris Trust.

Headland Archaeology Ltd. with John Renshaw Architects (2007) *The North Harris Trust Bunavoneader Whaling Station, Isle of Harris. Conservation and Management Plan.*

Heynen, N., McCarthy, J., Prudham, S. and Robbins, P. (eds) (2007a) *Neoliberal Environments. False Promises and Unnatural Consequences*, Routledge, London.

Heynen, N., McCarthy, J., Prudham, S. and Robbins, P. (2007b) Introduction: false promises, in *Neoliberal Environments. False Promises and Unnatural Consequences* (eds N. Heynen, J. McCarthy, S. Prudham and P. Robbins), Routledge, London, pp. 1–21.

Highlands and Islands Community Energy Company (HICEC) (2006). *Annual Review 2005–2006*, Highlands and Islands Enterprise, Inverness.

Hinchliffe, S. (2007) *Geographies of Nature. Societies, Environments, Ecologies*, Sage, London.

Hinrichs, C.C. (2003) The practice and politics of food system localization. *Journal of Rural Studies*, 19, 33–45.

Horner and MacLennan (2006) *Landscape and Visual Impact Assessment. North Harris Turbines*, report for the North Harris Trust.

Horner and MacLennan (2007) *Proposed Wind Turbines at Monan, Isle of Harris. Landscape and Visual Impacts*, report for the North Harris Trust.

Hunter, James (1974) The politics of Highland Land Reform, 1873–1895, *Scottish Historical Review*, 53 (1): 45–68.

Hunter, James (1976) *The Making of the Crofting Community*, John Donald, Edinburgh.

Hunter, James (1991) *The Claim of Crofting*, Mainstream, Edinburgh.

Hunter, James (1995a) *On the Other Side of Sorrow. Nature and People in the Scottish Highlands*, Mainstream, Edinburgh.

Hunter, James (1995b) Towards a land reform agenda for a Scots Parliament. *The Second John McEwen Memorial Lecture 1995*, Rural Forum.

Hunter, James (2009) *Keynote address to the Community Land Conference*, Tairbeart, Harris, 2009.

Hunter, James (2012) *From the Low Tide of the Sea to the Highest Mountain Tops*, The Islands Book Trust, Ravenspoint, Kershader, Isle of Lewis.

Hunter, Janet (2007) *A Future for North Harris: The North Harris Trust*, The North Harris Trust, Tarbert.

Hutchinson, R. (2003) *The Soap Man. Lewis, Harris and Lord Leverhulme*, Birlinn, Edinburgh.

Jay, M. (1994) *Downcast Eyes*, University of California Press, Berkeley, CA.

Jedrej, C. and Nuttall, M. (1996) *White Settlers: The Impact of Rural Repopulation in Scotland*, Harwood, Luxembourg.

John Muir Trust (2004) *Wild Land Policy.* www.jmt.org/policy-wild-lad.asp (accessed 30 August 2010).

Johnston, J., Biro, A. and MacKendrick, N. (2009) Lost in the supermarket: the corporate-organic foodscape and the struggle for food democracy. *Antipode*, 41 (3), 509–532.

Jones, G. (2011) *Trends in Common Grazing. First Steps Towards an Integrated Needs-Based Strategy*, European Forum on Nature Conservation and Pastoralism.

Katz, C. (1998) Whose nature, whose culture?: private productions of space and the "preservation" of nature, in *Remaking Reality* (eds B. Braun and N. Castree), Routledge, London, pp. 46–63.

Katz, C. (2005) Partners in crime? Neoliberalism and the production of new political subjectivities. *Antipode*, 37, 623–631.

Klein, N. (2001) Reclaiming the commons. *New Left Review*, 9, 81–89.

Knight Frank (2002) *The Amhuinnsuidhe and North Harris Estate, Isle of Harris, Outer Hebrides*, sales brochure, Knight Frank, Edinburgh.

Knott, C. (2003) *Archaeological Desk Based Assessment and Walk-Over Survey, North Harris, Western Isles*, commissioned by Edmund Nuttall Civil Engineering Ltd, in accordance with a standard brief prepared by Comhairle nan Eilean Siar.

Knott, C. and the Harris Archaeology Group (2006) *Archaeological Walk-over Survey. Monan Community Windfarm*, for the North Harris Trading Company.

Knoydart Foundation (2006) *Knoydart Foundation Housing Policy. Allocations Policy*.

Kohn, T. (2002) Becoming an islander through action in the Scottish Hebrides. *Journal of the Royal Anthropological Institute NS*, 8, 143–158.

Land Reform Policy Group (LRPG) (1998a) *Identifying the Problems*, The Scottish Office, Edinburgh.

Land Reform Policy Group (LRPG) (1998b) *Identifying the Solutions*, The Scottish Office, Edinburgh.

Land Reform Policy Group (LRPG) (1999) *Recommendations for Action*, The Scottish Office, Edinburgh.

Land Reform (Scotland) Act 2003 (2003) asp 2, Stationery Office, Edinburgh.

Law Society of Scotland (1987–1993) *The Laws of Scotland: Stair Memorial Encyclopaedia*, XVIII, Edinburgh.

Lawson, B. (2002) *Harris in History and Legend*, Donald, Edinburgh.

Leneman, L. (1989) *Fit for Heroes? Land Settlement in Scotland after World War I*, Aberdeen University Press, Aberdeen.

Lewis Wind Power (LWP) (2004) *Non-Technical Summary of the Environmental Statement*.

LHHP Newsletter (2005) November, 3.

LHHP Newsletter (2007) January, 7.

Logie, D. (2007) *Houses on Crofting Land. A Study Into Meeting Housing Needs in the Crofting Areas*, report by the Rural Housing Service for the Scottish Crofting Foundation.

Lorimer, H. (2000) Guns, game and the grandee: the cultural politics of deer-stalking in the Scottish Highlands. *Ecumene*, 7, 403–431.

MacAskill, J. (1999) *We Have Won The Land: The Story of the Purchase of the North Lochinver Estate*, Acair, Stornoway.

MacAskill, J. (2004) The crofting community right to buy in the Land Reform (Scotland) Act 2003. *Scottish Affairs*, 49, 104–133.

MacAskill, J. (2006) "The most arbitrary, scandalous act of tyranny": the Crown, private proprietors, and the ownership of the Scottish foreshore in the nineteenth century. *Scottish Historical Review*, 85 (2), 277–304.

MacAulay, J. (1996) *Birlinn: Longships of the Hebrides*, White Horse Press, Isle of Harris.

MacCormick, N. (1998) The English Constitution, the British State and the Scottish Anomaly. *Scottish Affairs*, Special Issue: Understanding Constitutional Change, 129–145.

MacDonald, C. (2010) Wind farms: pragmatic delivery of community benefit offers best way forward, Letter to the Editor. *West Highland Free Press*, 15 January, 17.

MacDonald, F. (1998) Viewing Highland Scotland: ideology, representation and the "natural heritage". *Area*, 30, 237–244.

MacDonald, F. (2001) St Kilda and the sublime. *Ecumene* 8 (2), 151–174.

Macdonald, S. (1997) *Reimagining Culture: Histories, Identities and the Gaelic Renaissance*, Berg, Oxford.

MacIlleathain, R. (2007) *A' Ghàidhlig air Aghaidh na Tìre: Ainmean-àite ann an Iar-thuath na Gàidhealtachd/Gaelic in the Landscape: Place Names in the North West Highlands*, Scottish National Heritage, Perth.

MacInnes, D. (2007) Support for wind farm proposals by 18–8 vote at the Comhairle. *Stornoway Gazette*, 22 February, 3.

MacInnes, D. (2008a) Island renewable groups share experiences at Uist Conference. *Stornoway Gazette*, 24 January, 7.

MacInnes, D. (2008b) "Historic day" as Beinn Mhor Power deal is signed. *Stornoway Gazette*, 31 July, 1.

MacInnes, D. (2008c) Islands' housing requirements at top of the agenda. *Stornoway Gazette*, 24 January, 11.

MacInnes, D. (2009a) Concern over "desecration of graves" at wind turbine site. *Stornoway Gazette*, 4 June, 1.

MacInnes, D. (2009b) Fewer, but taller turbines planned for Eishken site. *Stornoway Gazette*, 23 July, 3.

MacInnes, D. (2009c) Amended plans for Eishken wind farm on table. *Stornoway Gazette*, 3 September, 1.

MacInnes, D. (2009d) Pairc Trust deny any secrecy on ballot. *Stornoway Gazette*, 17 December, 4.

MacInnes, D. (2010) Grand vision for cash raised from Eishken wind farm. *Stornoway Gazette*, 21 January, 1.

MacInnes, J. (2006) *Dùthchas Nan Gàidheal. Selected Essays of John MacInnes* (ed. M. Newton), Birlinn, Edinburgh.

MacKay, D. (2007) Quangos "The New Landowners": their day is not yet done, Letter to the Editor. *West Highland Free Press*, 7 September, 13.

Mackay, K. (2007) North Harris needs support now, not at some time in the future, Letter to the Editor, *West Highland Free Press*, 20 July, 15.

Mackenzie, A.F.D. (1998a) *Land, Ecology and Resistance in Kenya, 1880–1952*, University of Edinburgh Press, Edinburgh, and Heinemann, USA.

Mackenzie, A.F.D. (1998b) "The Cheviot, The Stag … and The White, White Rock?": community, identity and environmental threat on the Isle of Harris. *Environment and Planning D: Society and Space*, 16, 509–532.

Mackenzie, A.F.D. (2001) On the edge: "community" and "sustainability" on the Isle of Harris, Outer Hebrides. *Scottish Geographical Journal*, 117 (3), 219–240.

Mackenzie, A.F.D. (2002) Re-claiming place: The Millennium Forest, Borgie, North Sutherland, Scotland. *Environment and Planning D: Society and Space*, 20, 535–560.

Mackenzie, A.F.D. (2006a) 'S Leinn Fhèin am Fearann (The Land is Ours): re-claiming land, re-creating community, North Harris, Outer Hebrides, Scotland. *Environment and Planning D: Society and Space*, 24, 577–598.

Mackenzie, A.F.D. (2006b) A working land: crofting communities, place and the politics of the possible in post-Land Reform Scotland. *Transactions of the Institute of British Geographers NS*, 31, 383–398.

Mackenzie, A.F.D. (2007) "Dismay" over inquiry into plan for Harris Community Wind Farm, Letter to the Editor. *West Highland Free Press*, 27 July, 15.

Mackenzie, A.F.D. (2009) Working the wind: land owning community trusts and the decolonisation of nature. *Scottish Affairs*, 66, 44–64.

Mackenzie, A.F.D. (2010a) Re-stor(y)ing north west Scotland. *Scottish Geographical Journal*, 126 (3), 162–184.

Mackenzie, A.F.D. (2010b) A common claim: community land ownership in the Outer Hebrides. *International Journal of the Commons*, 4 (1), 319–344.

MacKenzie, J. (2003) Assynt Hydro Scheme, in *The Next Generation*, Report of the 'Fling the Fank' Conference, Assynt Crofters' Trust, Stoer, North Assynt, 28–30 August 2003: 3.

Mackenzie, K. (2007) Kilvaxter – a symptom of a wider decline in the crofting way of life. *West Highland Free Press*, 22 June, 11.

Mackinnon, I. (2005) The link between holiday homes and the lack of affordable housing. *West Highland Free Press*, 14 October, 12.

Maclean, S. (1941) The poetry of the Clearances. *Transactions of the Gaelic Society of Inverness*, 38, 293–324.

Maclennan, J. (2001) *Place-names of Scarp*, Stornoway Gazette, Stornoway.

Maclennan, K. (2011) *Final Powerdown Report*, Galson Estate Trust.

Macleod, C., Braunholtz-Speight, T., MacPhail, I., Flyn, D., Allen, S. and Macleod, D. (2010) *Post Legislative Scrutiny of the Land Reform (Scotland) Act 2003*. Final Report, Scottish Government, Edinburgh.

Macleod, M. (2009) Affordable homes for locals enjoying an Eigg roll. *The Times*, 25 April, 36.

Macleod, M. and Wade, M. (2009) Rum deal as island's fate rests with seventeen voters. *The Times*, 13 January, 21.

MacMillan, D.C., Leitch, K., Wightman, A. and Higgins, P. (2010) The management role of Highland sporting estates in the early twenty-first century: the owner's view of a unique but contested form of land use. *Scottish Geographical Journal*, 126 (1), 24–40.

Macpherson Research (2003) *Western Isles Tourism Facts and Figures*, Update 1 (this information was supplied to CnES, Stornoway).

MacSween, I. (2002) Hearaich urged to grasp the chance to become their own landlords. *Stornoway Gazette*, 2 May, 1, 3.

MacSween, I. (2003) North Harris – it's our land. *Stornoway Gazette*, 6 March, 1, 2.

MacSween, I. (2005) Vice Convener slams "childish and patronizing RSPB". *Stornoway Gazette*, 1 December, 2.

Mansfield, B. (2007a) Privatization: property and the remaking of nature–society relations. *Antipode*, 39 (3), 393–405.

Mansfield, B. (2007b) Neoliberalism in the oceans. "Rationalization," property rights, and the commons question, in *Neoliberal Environments. False Promises and Unnatural Consequences* (eds N. Heynen, J. McCarthy, S. Prudham and P. Robbins), Routledge, Abingdon, pp. 63–73.

Market Research Partners, Edinburgh (2008) Public perceptions of wild places and landscapes in Scotland. *Commissioned Report 291 for Scottish Natural Heritage*, ROAME No. F06NC03.

Martin, P. and Chang, X. (2007) Bere and beer. *The Brewer and Distiller International*, 3 (6), 29.

Massey, D. (2000) Entanglements of power. Reflections, in *Entanglements of Power. Geographies of Domination/Resistance* (eds J.P. Sharp, P. Routledge, C. Philo and R. Paddison), Routledge, London, pp. 279–286.

Massey, D. (2005) *For Space*, Sage, London.

Massey, D., with the Human Geography Research Group, University of Glasgow, eds S. Bond and D. Featherstone (2009) The possibilities of a politics of place beyond place? A conversation with Doreen Massey. *Scottish Geographical Journal*, 125 (3/4), 401–420.

Mather, A.S. (1993) Protected areas in the periphery: conservation and controversy in Northern Scotland. *Journal of Rural Studies*, 9, 371–384.

McCarthy, J. (2005a) Commons as counterhegemonic projects. *Capitalism Nature Socialism*, 16, 9–24.

McCarthy, J. (2005b) Devolution in the woods: community forestry as hybrid neoliberalism. *Environment and Planning A*, 37, 995–1014.

McCarthy, J. and Prudham, S. (2004) Neoliberal nature and the nature of neoliberalism. *Geoforum*, 35, 275–283.

McCrone, D. (1997) *Land, Democracy and Culture in Scotland*, The Fourth McEwen Lecture on Land Tenure in Scotland, A K Bell Library, Perth.

McDade, H. (2010) Making a stand. Learning the lessons from Beauly–Denny. *John Muir Trust Journal*, 48, 16–18.

McGrath, J. (1981) *The Cheviot, The Stag, and the Black, Black Oil*, Eyre Methuen, London.

McHardy, I. (2006a) Blighting a valuable landscape, Letter to the Editor. *Stornoway Gazette*, 29 June, 8.

McHardy, I. (2006b) *The Archaeology of an Area of Langadale, North Harris*, carried out for the North Harris Trust.

McHardy, I. (2008) Summation of evidence given by Mr Ian McCardy at the Public Local Inquiry into the proposed development of a windfarm on the Eisgen Estate, Isle of Lewis, by Beinn Mhor Power, held from the 15th to the 22nd of May 2008, Stornoway, Isle of Lewis. Available at Comhairle nan Eilean Siar, Stornoway.

McHardy, I. (2010) *Callanish – Monument, Moon and Mountain*, The Islands Book Trust, Kershader, Isle of Lewis.

McIntosh, A. (2004) *Soil and Soul. People Versus Corporate Power*, Aurum, London.

McIntosh, A. (2009) Time to build on what crofting has and make it work properly, Letter to the Editor. *West Highland Free Press*, 23 October, 17.

McMorran, R., Price, M.F. and McVittie, A. (2006) A review of the benefits and opportunities attributed to Scotland's landscapes of wild character, *Scottish Natural Heritage Commissioned Report 194*, ROAME No. F04NC18.

Meek, D.E. (1976) Gaelic poets of the Land Agitation, *Transactions of the Gaelic Society of Inverness*, 309–376.

Meyer, N.I. (2007) Learning from wind energy policy in the EU: lessons from Denmark, Sweden and Spain. *European Environment*, 17 (5), 347–362.

Moore, D. S., Pandian, A. and Kosek, J. (2003) Introduction. The cultural politics of race and nature: terrains of power and practice, in *Race, Nature and the Politics of Difference* (eds D.S. Moore, A. Pandian and J. Kosek), Duke University Press, Durham, NC), pp. 1–70.

Morrison, J. (2006) A God's island and the Scottish Land Fund. *West Highland Free Press*, 3 November, 10.

Mudimbe, V.Y. (1988) *The Invention of Africa: Gnosis, Philosophy, and the Order of Knowledge*, Indiana University Press, Bloomington, IN.

Murray, R., Maclean, M. and Gordon, L. (1995) *Calanais*, An Lanntair, Stornoway.

Nancy, J.-L. (1991) Of being-in-common, in *Community at Loose Ends* (ed. The Miami Theory Collective), University of Minnesota Press, Minneapolis, MN, pp. 1–12.

Nash, C. (2002) Genealogical identities. *Environment and Planning D: Society and Space*, 20, 27–52.

National Parks (Scotland) Act 2000 (2000) asp 10, Stationary Office, Edinburgh.

National Trust for Scotland (2002) *Wild Land Policy*, http://www.nts.org.uk/conserve/downloads/wild_land_policy_2002.pdf (last accessed 28 March 2012).

Newton, M. (2009) *Warriors of the Word. The World of the Scottish Highlanders*, Edinburgh, Birlinn.

Nic Bheatha/Beith, M. (2000) aibheil na craobh ogam/Gaelic tree alphabet, in *A'Chraobh/The Tree*, Dornoch Studio, Dornoch, pp. 17–45.

North Harris Steering Group (2002) *Steering Group's Recommendation Presentation at Public Meeting*, presented by David Cameron.

North Harris Trust (NHT) (2002) *Memorandum and Articles of Association.*
North Harris Trust (NHT) (2003) *Baseline Study.*
North Harris Trust (NHT) (2006) *Bye-Laws of the North Harris Trust.*
North Harris Trust (NHT) (2007) *Land Management Plan*, Tarbert, Harris.
North Harris Trust (NHT) (2008) News release, 12 March.
North Harris Trust (NHT) (2009a) *North Harris Deer Management Plan 2009–2012*, Tarbert, Harris.
North Harris Trust (NHT) (2009b) *North Harris Trust Land Release Policy*, Tarbert, Harris.
North Harris Trust Land Manager's report (2010) *Dè Tha Dol?* 12 March: 24.
O'Drisceoil, S. (2009) *Lewis and Harris Rural Community Housing Pilot Project*, Comhairle nan Eilean Siar, Stornoway.
Ostrom, E. (1990) *Governing the Commons: The Evolution of Institutions for Collective Action*, Cambridge University Press, Cambridge.
Ostrom, E. (2005) *Understanding Institutional Diversity*, Princeton University Press, Princeton, NJ.
Ostrom, E., Burger, J., Field, C.B., Norgaard, R.B. and Policansky, D. (1999) Revisiting the commons: local lessons, global challenges. *Science*, 284, 278–282.
Owens, S. and R. Cowell (1996) *Rocks and Hard Places. Mineral Resource Planning and Sustainability*, report for the Council for the Protection of Rural England, London.
Pairc Trust (2009) *Pairc Community Asks Ministers to Help Overcome Landlord Delaying Tactics*, press release, 30 September.
Parrott, J. and MacKenzie, N. (2000) *Restoring and Managing Riparian Woodlands*, Scottish Native Woods, Aberfeldy.
Pasqualetti, M.J., Gipe, P. and Richter, R.W. (2002) A landscape of power, in *Wind Power in View: Energy Landscapes in a Crowded World* (eds M.J. Pasqualetti, P. Gipe and R.W. Richter), Academic, San Diego, CA, pp. 3–16.
Peck, J., Theodore, N. and Brenner, N. (2010) Postneoliberalism and its malcontents, in *The Point Is To Change It* (eds N. Castree, P. Chatterton, N. Heynen, W. Larner and M.W. Wright), Wiley-Blackwell, Chichester, pp. 94–116.
Peck, J. and Tickell, A. (2002) Neoliberalizing space. *Antipode*, 34 (3), 380–404.
Peet, R. and Watts, M. (eds) (2004) *Liberation Ecologies: Environment, Development, Social Movements*, Routledge, London.
Phillips, J. (2005) Review of Lindsay, R.A. and O.M. Bragg, 2004, "Wind Farms and Blanket Mires: the Bog Slide of 16 October 2003 at Derrybrien, Co. Galway". *The Views of Scotland Newsletter*, 2 (4), 8–10.
Pillai, A. (2012) Land law, draft chapter prepared for *Scottish Life and Society: A Compendium of Scottish Ethnography. Vol. 13: Institutions of Scotland – The Law* (ed. M. Mulhern), University of Edinburgh Press, Edinburgh, in press.
Press and Journal (2009) 23 April: 6.
Prudham, S. (2007) The fictions of autonomous invention: accumulation by dispossession, commodification and life patents in Canada. *Antipode*, 39 (3), 406–429.

Rajchman, J. (1991) *Philosophical Events. Essays of the '80s*, Columbia University Press, New York.

Randall, J. (2010) Experience of RTB [Right To Buy] – community perspectives and experience of RTB, legislation and process, in *Land Reform. Rights to Buy – Where To from Here?*, Conference Report, Inverness, 2010, pp. 18–19.

Reid, D. (2003) Crofters common grazings. *Commonweal of Scotland Working Paper* 2, http://www.scottishcommons.org/docs/commonweal_2.pdf (last accessed 28 March 2012).

Reid, K. and van der Merwe, C.G. (2004) Property law: some themes and some variations, in *Mixed Legal Systems in Comparative Perspective: Property and Obligations in Scotland and South Africa* (eds K. Reid and C.G. van der Merwe), Oxford University Press, Oxford, pp. 638–670.

Reid, R. (2008) *Golden Eagle Diet and Breeding Success on the Western Isles, Scotland*, degree project in Animal Ecology, Department of Animal Ecology, Lund University.

Rennie, Agnes (2010) Land reform – our legacy from generations past, creating opportunities for future generations. *Seventh Angus Macleod Memorial Lecture*, Islands Book Trust, Kershader, South Lochs, Isle of Lewis.

Rennie, Alison (2006) The importance of national parks to nation-building: support for The National Parks Act (2000) in the Scottish Parliament. *Scottish Geographical Journal*, 122 (3), 223–232.

Riddington, G., Harrison, T., McArthur, D., Gibson, H. and Millar, K. (2008) *The Economic Impacts of Wind Farms on Scottish Tourism*, a report for the Scottish Government.

Robbins, P. and Fraser, A. (2003) A forest of contradictions: producing the landscapes of the Scottish Highlands. *Antipode*, 35, 95–118.

Robertson, I.J.M. (1997) The role of women in social protest in the Highlands of Scotland. *Journal of Historical Geography*, 23 (2), 187–200.

Robertson, M.M. (2007) The neoliberalization of ecosystem services: wetland mitigation banking and the problem of measurement, in *Neoliberal Environments* (eds N. Heynen, J. McCarthy, S. Prudham and P. Robbins), Routledge, London, pp. 114–125.

Rodway, P. (2009) The launch of Crofting Connections. *The Crofter*, 85, 16.

Ross, D. (2009) Plan to open new St. Kilda centre for visitors. *The Herald*, 23 April, 12.

Royal Society for the Protection of Birds Scotland (2005) *Strategical Locational Guidance for Wind Farm Developments*.

Rural Housing Service (2002) *Isle of Gigha Housing Needs Study*, a report to Isle of Gigha Heritage Trust, Argyll and the Isles Enterprise and Communities Scotland.

Russell, M. (2008) Bringing power to the people. *West Highland Free Press*, 27 June, 11.

Said, E. (1994) *Culture and Imperialism*, Knopf, New York.

Satsangi, M. (2007) Land tenure change and rural housing in Scotland. *Scottish Geographical Journal*, 123 (1), 33–47.

Scholten, M. (2008) A wee note on coerce beag – with a riddle about stooks. *West Highland Free Press*, 5 December, 18.

Scholten, M. (2009) Farmers' rights – what's in it for crofters? *The Crofter*, March, 17.

Scholten, M., Maxted, N., Ford-Lloyd, B.V. and Green, N. (2008) Hebridean and Shetland oat (*Avena strigosa Schreb.*) and Shetland Cabbage (*Brassica oleracea L.*) landraces: occurrence and conservation issues. *PGR Newsletter* 154, 1–8.

Scholten, M., Spoor, B., Carter, S., MacPherson, N. (2010) Genetic Resources and Plant Evolution or Conservation Biology (unpublished paper).

Scott, J.C. (1998) *Seeing Like a State: How Certain Schemes to Improve the Human Condition Have Failed*, Yale University Press, New Haven, CT.

Scott, N.C. (2007) Grid connection of the Scottish Islands. A strategic viewpoint. *Report 1007/001/001 D commissioned by Highlands and Islands Enterprise, Orkney Islands Council, Shetland Islands Council and Comhairle nan Eilean Siar*, Xero Energy, Glasgow.

Scottish Executive (1999a) *NPPG 14: Natural Heritage*, http://scotland.gov.uk/Publications/1999/01/nppg14 (last accessed 28 March 2012).

Scottish Executive (1999b) *Land Reform: Proposals for Legislation*, Scottish Executive, Edinburgh.

Scottish Executive (2004) *Scotland's Biodiversity: It's In Your Hands*, Scottish Executive, Edinburgh.

Scottish Government (2008a) *Framework for the Development and Deployment of Renewables in Scotland*, Scottish Government, Edinburgh.

Scottish Government (2008b) *Climate Change: Consultation on Proposals for a Scottish Climate Change Bill*, Scottish Government, Edinburgh.

Scottish Government (2008c) *Committee of Inquiry on Crofting. Government Response*, Scottish Government, Edinburgh.

Scottish Government (2009) *Community. Scottish Community Empowerment Action Plan*, Scottish Government, Edinburgh.

Scottish Government (2010) *Speak Up for Rural Scotland*, Scottish Government, Edinburgh.

Scottish Natural Heritage (SNH) (2001) *Guidelines on the Environmental Impact of Windfarms and Small Scale Hydroelectric Schemes*, SNH, Redgorton, Perth.

Scottish Natural Heritage (SNH) (2002) Wildness in Scotland's countryside. *Policy Statement 02/03*, SNH, Redgorton, Perth.

Scottish Natural Heritage (SNH) (2004) *Native Woodland Model*, SNH, Redgorton, Perth.

Scottish Natural Heritage (SNH) (2005a) Landscape policy framework. *Policy Statement 05/01*, SNH, Inverness.

Scottish Natural Heritage (SNH) (2005b) Strategic locational guidance for onshore wind farms in respect of the natural heritage. *Policy Statement 02/02*, update May, SNH, Inverness.

Scottish Natural Heritage (SNH) (2007) *Section 15 Agreement between North Harris Trust and Scottish Natural Heritage*, SNH, Stornoway.

Scottish Natural Heritage (SNH) (2008a) *Guidance for Identifying the Special Qualities of Scotland's National Scenic Areas*, final version 29 January, SNH, Inverness.

Scottish Natural Heritage (SNH) (2008b) *Western Isles Native Woodland Restoration Survey Report*, SNH and Comhairle nan Elean Siar, Stornoway.

Seller, J. (2006) Freshwater pearl mussel translocation – River -------, North Harris SAC, Western Isles. *Final Report A2406/R1/Rev1*, prepared for Scottish Natural Heritage, Young Associates.

Sime, I. (2003) *River Runners – a Tale of Protected Species*, Scottish Natural Heritage, Redgorton, Perth.

Singer, J.W. (2000) Property and social relations: from title to entitlement, in *Property and Values: Alternatives to Public and Private Ownership* (eds C. Geisler and G. Daneker), Island, Washington, DC, pp. 3–20.

Singer, L. (1991) Recalling a community at loose ends, in *Community at Loose Ends* (ed. Miami Theory Collective), University of Minnesota Press, Minneapolis, MN, pp. 121–130.

Skerratt, S. (2011) Community land ownership and community resilience. *Rural Policy Centre Research Report*, Scottish Agricultural College, http://www.sac.ac.uk/research/groups/lee/teams/ruralsociety/ (last accessed 28 March 2012).

Skinner, A., Young, M. and Hastie, L. (2003) Ecology of the freshwater pearl mussel. *Conserving Natura 2000 Rivers Ecology Series 2*, English Nature, Peterborough.

Smith, A. (2009) Battle site unknown. *Stornoway Gazette*, 11 June, 4.

Smith, N. (2008) *Uneven Development: Nature, Capital, and the Production of Space*, University of Georgia Press, Athens, GA.

Smout, C., MacDonald, A.R. and Watson, F. (eds) (2005) *A History of the Native Woodlands of Scotland, 1500–1920*, Edinburgh University Press, Edinburgh.

Spencer, F., for the National Parks Review Team (2008) *National Parks Strategic Review Report*, commissioned by the Minister for the Environment, Michael Russell, http://www.scotland.gov.uk/Topics/Environment/Countryside/16131 (last accessed 28 March 2012).

SQW (2005) *Evaluation of the Community Land Unit*, a final report to Highlands and Islands Enterprise.

St. Martin, K. (2007) Enclosure and economic identity in New England fisheries, in *Neoliberal Environments: False Promises and Unnatural Consequences* (eds N. Heynen, J. McCarthy, S. Prudham and P. Robbins), Routledge, New York, pp. 255–268.

Steven, A.J.M. (2004) Revolution in Scottish Land Law. *Electronic Journal of Comparative Law*, 8 (3), 1–13.

Stockdale, A. (2010) The diverse geographies of rural gentrification in Scotland. *Journal of Rural Studies*, 26, 31–40.

Stockdale, A. and Barker, A. (2009) Sustainability and the multifunctional landscape: an assessment of approaches to planning and management in the Cairngorms National Park. *Land Use Policy*, 26, 479–492.

Stornoway Gazette (2001) 14 June, 3.

Stornoway Gazette (2004) 8 January, 3.

Stornoway Gazette (2004) 19 February, 3.

Stornoway Gazette (2005) 23 June, 3.

Stornoway Gazette (2005) 30 June, 4.

Stornoway Gazette (2005) 17 July, 3.

Stornoway Gazette (2005) 15 September, 14.

Stornoway Gazette (2005) 6 October, 3.

Stornoway Gazette (2005) 22 December, 5.

Stornoway Gazette (2006) 16 February, 19.

Stornoway Gazette (2006) 9 March, 5.

Stornoway Gazette (2008) 21 April, 11, 12.

Stornoway Gazette (2009) 9 April, 3.

Stornoway Gazette (2009) 23 April, 18.

Stornoway Gazette (2009) 21 May, 3.

Stornoway Gazette (2009) 21 May, 7.

Sweeney, C. (2009) Paradise comes at a price for island's 151 owners. *The Times*, 30 May, 26.

Swyngedow, E. (2005) Dispossessing H_2O: the contested terrain of water privatization. *Capitalism Nature Socialism*, 16 (1), 81–98.

Theobald, H.E., Wishart, J.E., Martin, P.J., Buttriss, J.L. and French, J.H. (2006) The nutritional properties of flours derived from Orkney grown bere barley (*Hordeum vulgare L.*). *British Nutrition Foundation Nutrition Bulletin*, 31, 8–14.

Thompson, N. (2005) The practice of government in a devolved Scotland: the case of the designation of the Cairngorms National Park. *Paper Presented at the Annual Conference of the Royal Geographical Society/Institute of British Geographers*, London, 2005.

Thomson, D. (1990) *An Introduction to Gaelic Poetry*, Edinburgh University Press, Edinburgh.

Tipping, R. (2005) Living in the past: woods and people in prehistory to 1000 BC, in *People and Woods in Scotland* (ed. T.C. Smout), Edinburgh University Press, Edinburgh.

TNEI Services (2007) *Assessment of the Grid Connection Options for the Scottish Islands*, report commissioned by Highlands and Islands Enterprise, TNEI Services, Manchester.

Toogood, M. (2003) Decolonizing highland conservation, in *Decolonizing Nature: Strategies for Conservation in a Post-colonial Era* (eds W.M. Adams and M. Mulligan), Earthscan, London, pp. 152–171.

Touchstone Heritage Management Consultants (THMC) and J. Scott (2006) *North Harris Interpretation Plan*, commissioned by the North Harris Trust.

UK Government (1884) *Evidence Taken by Her Majesty's Commissioners of Inquiry into the Condition of the Crofters and Cottars in the Highlands and Islands of Scotland*, Vol. 1, Neill, Edinburgh.

van der Horst, D. and Vermeylen, S. (2008) The new energy commons: exploring the role of property regimes in the development of renewable energy systems. *Paper Presented at the Conference of the International Association for the Study of Commons*, Cheltenham.

Walker, G. and Cass, N. (2007) Carbon reduction, "the public" and renewable energy: engaging with socio-technical configurations. *Area*, 39 (4), 458–469.

Walker, G. and Devine-Wright, P. (2008) Community renewable energy: what should it mean? *Energy Policy*, 36 (2), 497–500.

Walker, G., Hunter, S., Devine-Wright, P., Evans, B. and Fay, H. (2007) Harnessing community energies: explaining and evaluating community-based localism in renewable energy policy in the UK. *Global Environmental Politics*, 7, 64–82.

Warren, C. (2007) Perspectives on the "alien" versus "native" species debate: a critique of concepts, language and practice. *Progress in Human Geography*, 31 (4), 427–446.

Warren, C. (2009) *Managing Scotland's Environment*, University of Edinburgh Press, Edinburgh.

Warren, C.R. and Birnie, R.V. (2009) Re-powering Scotland: wind farms and the "energy or environment?" debate. *Scottish Geographical Journal*, 125 (2), 97–126.

Warren, C.R. and McFadyen, M. (2010) Does community ownership affect public attitudes to wind energy? A case study from south-west Scotland. *Land Use Policy*, 27, 204–213.

Watts, M. (2004) Antimonies of community: some thoughts on geography, resources and empire. *Transactions of the Institute of British Geographers NS*, 29, 195–216.

Watts, M. (2007) What might resistance to neoliberalism consist of? in *Neoliberal Environments* (eds N. Heynen, J. McCarthy, S. Prudham and P. Robbins), Routledge, London, pp. 273–278.

West Coast Energy (2004) *Study into the Possibilities for Renewable Energy Developments in North Harris*, Tower Mains Studios, Edinburgh.

West Highland Free Press (WHFP) (2002) 17 May, 1.
West Highland Free Press (WHFP) (2002) 20 September, 1.
West Highland Free Press (WHFP) (2003) 11 July, 3.
West Highland Free Press (WHFP) (2003) 22 August, 9.
West Highland Free Press (WHFP) (2003) 29 August, 2.
West Highland Free Press (WHFP) (2003) 26 December, 3.
West Highland Free Press (WHFP) (2004) 9 April, 13.
West Highland Free Press (WHFP) (2004) 3 September, 3.
West Highland Free Press (WHFP) (2004) 10 September, 2.
West Highland Free Press (WHFP) (2004) 3 December, 1.
West Highland Free Press (WHFP) (2005) 21 January, 5.
West Highland Free Press (WHFP) (2005) 18 February, 3.

West Highland Free Press (*WHFP*) (2005) 18 March, 11.
West Highland Free Press (*WHFP*) (2005) 3 June, 1.
West Highland Free Press (*WHFP*) (2005) 24 June, 5.
West Highland Free Press (*WHFP*) (2005) 30 September, 3.
West Highland Free Press (*WHFP*) (2005) 21 October, 1, 3.
West Highland Free Press (*WHFP*) (2005) Editorial. 21 October, 17.
West Highland Free Press (*WHFP*) (2005) Editorial. 4 November, 7.
West Highland Free Press (*WHFP*) (2006) 10 February, 3.
West Highland Free Press (*WHFP*) (2006) 10 March, 5.
West Highland Free Press (*WHFP*) (2006) 26 May, 5.
West Highland Free Press (*WHFP*) (2006) 22 September, 3.
West Highland Free Press (*WHFP*) (2006) 22 December, 11.
West Highland Free Press (*WHFP*) (2007) 12 January, 3.
West Highland Free Press (*WHFP*) (2007) 19 January, 1.
West Highland Free Press (*WHFP*) (2007) 26 January, 2.
West Highland Free Press (*WHFP*) (2007) Editorial. 26 January, 13.
West Highland Free Press (*WHFP*) (2007), 2 February, 11.
West Highland Free Press (*WHFP*) (2007) 16 February, 5.
West Highland Free Press (*WHFP*) (2007) 23 February, 7.
West Highland Free Press (*WHFP*) (2007) 4 May, 7.
West Highland Free Press (*WHFP*) (2007) 18 May, 19.
West Highland Free Press (*WHFP*) (2007) 1 June, 2.
West Highland Free Press (*WHFP*) (2007) 6 July, 3.
West Highland Free Press (*WHFP*) (2007) Editorial. 20 July, 15.
West Highland Free Press (*WHFP*) (2007) 5 October, 3.
West Highland Free Press (*WHFP*) (2007) 30 November, 9.
West Highland Free Press (*WHFP*) (2008) 9 January, 1.
West Highland Free Press (*WHFP*) (2008) 8 February, 1.
West Highland Free Press (*WHFP*) (2008) 22 February, 2.
West Highland Free Press (*WHFP*) (2008) 29 February, 13.
West Highland Free Press (*WHFP*) (2008) 18 April, 9.
West Highland Free Press (*WHFP*) (2008) 13 June, 2.
West Highland Free Press (*WHFP*) (2008) 21 August, 3.
West Highland Free Press (*WHFP*) (2008) 29 August, 22, 23.
West Highland Free Press (*WHFP*) (2008) 10 October, 2.
West Highland Free Press (*WHFP*) (2008) 17 October, 9.
West Highland Free Press (*WHFP*) (2009) 2 January, 3.
West Highland Free Press (*WHFP*) (2009) 27 February, 3.
West Highland Free Press (*WHFP*) (2009) 6 March, 2.
West Highland Free Press (*WHFP*) (2009) 6 March, 15.
West Highland Free Press (*WHFP*) (2009) 8 May, 1.
West Highland Free Press (*WHFP*) (2009) 10 August, 1, 3.
West Highland Free Press (*WHFP*) (2009) Editorial. 6 November, 15.
West Highland Free Press (*WHFP*) (2009) 25 December, 1.

West Highland Free Press (*WHFP*) (2010) 22 January, 3.
West Highland Free Press (*WHFP*) (2010) 5 March, 1, 3.
West Highland Free Press (*WHFP*) (2010) 7 May, 1, 2.
West Highland Free Press (*WHFP*) (2010) 2 July, 3.
West Highland Free Press (*WHFP*) (2010) 8 October, 3.
West Highland Free Press (*WHFP*) (2010) 10 October, 1.
West Highland Free Press (*WHFP*) (2011) 7 January, 1.
West Highland Free Press (*WHFP*) (2011) Editorial. 28 January, 15.
West Highland Free Press (*WHFP*) (2011) 22 February, 2.
West Highland Free Press (*WHFP*) (2011) 18 March, 1.
Western Isles Development Trust (2005) *Business Plan 2005–2008*.
Whatmore, S. (2002) *Hybrid Geographies: Natures, Cultures, Spaces*, Sage, London.
Wightman, A. (1996) *Who Owns Scotland*, Canongate, Edinburgh.
Wightman, A. (2007) *Land Reform (Scotland) Act 2003: (Part 2, the Community Right to Buy) a Two-Year Review*, Caledonia Centre for Social Development, Inverness.
Wightman, A. (2010) *The Poor Had No Lawyers. Who Owns Scotland (And How They Got It)*, Birlinn, Edinburgh.
Wightman, A. (2011) *Land Reform. The Way Ahead*, Scottish Community Alliance, Edinburgh.
Withers, C.W.J. (1988) *Gaelic Scotland: The Transformation of a Culture Region*, Routledge, London.
Withers, C.W.J. (1996) Place, memory, monument: memorializing the past in contemporary Highland Scotland, *Ecumene*, 3, 325–344.
Wittman, H. (2009) Reworking the metabolic rift: La Vía Campesina, agrarian citizenship, and food sovereignty. *Journal of Peasant Studies*, 36 (4), 805–826.
Wolford, W. (2007) Neoliberalism and the struggle for land in Brazil, in *Neoliberal Environments* (eds N. Heynen, J. McCarthy, S. Prudham and P. Robbins), Routledge, London, pp. 243–254.
Woods, M. (2003) Deconstructing rural protest: the emergence of a new social movement. *Journal of Rural Studies*, 19, 309–325.
Young (Solicitors). (2004) *Report for the Isle of Gigha Heritage Trust. Subject: Protecting the Trust's Interests in the Sale of Property on Gigha*, Edinburgh.

Index

Places of Possibility: Property, Nature and Community Land Ownership,
First Edition. A. Fiona D. Mackenzie.
© 2013 A. Fiona D. Mackenzie. Published 2013 by Blackwell Publishing Ltd.